AF240874

The Lost Art of Perfume

Septimus Piesse

The Lost Art of Perfume

Max Milo
ESSAIS-DOCUMENTS

Max Milo Éditions, Paris, 2023
www.maxmilo.com
ISBN : 978-2-31501-231-2

THE LOST ART OF PERFUME

I. PERFUMERY THROUGH THE CENTURIES

The Creator's hand has poured all the treasures of his wealth onto flowers; after placing them on graceful, delicate stems, he has painted them with the most vivid, varied and harmonious colors, and impregnated them with the most exquisite scents. The flower occupies an important place in nature's designs. In the middle of the corolla are placed the reproductive organs destined to perpetuate the species after the death of the individuals who have borne them; so, from the cradle, man has learned to respect such magnificent productions; and if he has sometimes allowed himself to put a reckless hand on them, it has undoubtedly only been to pay homage to the Creator of what he considered to be the symbol of perfection.

In recognition of God's sovereign dominion over his creatures, man has sought to extract the sweet perfumes they contain from flowers, and has taken from various plants the intoxicating resins and balms, some of which are still burned in our temples.

The first traces of the perfumer's art can be found in the religious ceremonies of the earliest peoples; among the nations of antiquity, an offering of perfumes was regarded as a token of the deepest and most respectful veneration.

Pliny places the origin of perfumery in those beautiful lands of the Orient, where plant riches are found together as in a privileged country which, being the first to receive the sun's rays, seems to exhaust all its invigorating virtues; his opinion is confirmed by the Holy Scriptures. The Bible's frequent references to perfumes and aromatics prove that, from very early on, they were widely consumed by peoples whose soil produced aloes, cinnamon, sandalwood, camphor, nutmeg and cloves, the incense tree, from which the Sabeans had the holy privilege of collecting the gum-resin, the *balsam* or *balsam tree*, the sad *Nyctanthes*, which spreads its rich perfumes at dusk, the *nilica*, in whose flowers the bees, it is said, fall asleep to the sound of their own buzzing ; all these plants and a host of others no less fragrant belong to the East, and for centuries remained unknown to the rest of the world.

1. Jewish Perfumes

That the ancients attached an idea not only of personal respect, but also of religious homage, to an offering of incense, is proved by the example of the Magi who, after prostrating themselves on their knees to adore the newborn Jesus and having recognized his divinity, offered him gold, myrrh and incense.

It doesn't seem that the Jews made much use of perfumes for their personal hygiene, as the strict prescriptions of the

The Lost Art of Perfume

Law of Moses against those who used perfumes reserved for the sanctuary may have deterred them from doing so. It is certain, however, that in addition to the perfumes burnt in the temple, the Jews had others which they poured over the dead. We see this in the example of Joseph of Arimathea, who anointed the body of Jesus before burying him, and in Jesus' reply to Judas, who reproached Mary for using a preparation borrowed from banquets as an unnecessary luxury: "She is preparing my burial. We find in St. John: "It was their custom to sprinkle aromatic substances on the dead, especially myrrh and aloes, which came from Arabia." This ceremony was expressed in ancient Greek by a verb meaning "to embalm" or "to bury"; it was performed by neighbors and relatives.

The Jews used to anoint themselves with perfumes before meals. However, it seems certain that they made little progress in the art of perfuming, and were content to use aromatics as nature offered them, or at most dissolved in suitable vehicles.

The early Christians imitated the Jews and adopted the use of incense in liturgical ceremonies. St. Ephraim, father of the Syrian Church, wrote in his will that no perfume should be burned or scattered on his coffin, but rather that the aromatics should be given to the sanctuary.

Perfumes were used in the service of the Church not only in the form of incense, but also mixed with oil and wax for the lamps and candles that were to burn in the Lord's house.

2. Chinese Perfumes

The Chinese, whose sensualism is so refined, make great use of perfumes, to which they give a large place in their worship, their domestic uses and their pleasures; they burn fragrant woods and resins before their altars and mix them with food; it is above all aphrodisiacs that are sought after ; and it is said that they know how to prepare certain scented balls made of amber, musk and hemp flowers mixed with opium and other more energetic substances; heated for a while and rolled in the hand, they throw the small-footed beauties who populate the Celestial Empire into a voluptuous spasm.

3. Perfumes in the Orient

Zoroaster's followers prayed before altars where the sacred fire shone, and five times a day the priests put wood and odors on them.

In the East, it is now considered a proof of friendship and an act of hospitality to sprinkle visitors with rose oil, or to perfume them with aloe wood at the end of each visit. In an excellent work that paints a vivid picture of Eastern domestic life, we find several passages on the use of perfumes. Such is the *story of the barber's younger brother*, who, finding himself lured into the palace of the Grand Vizier's wife to become her plaything and amusement, saw himself "painted like a woman, shaved and then perfumed with aloe wood and rosewater".

4. Scythian Perfumes

Herodotus tells us that Scythian women used to grind cypress wood, cedar and frankincense on a stone; they then poured in a certain amount of water until the mixture took on the consistency of a paste that was used to coat the face and limbs; this composition first gave off a pleasant smell, then when it was removed the next day, it made the skin soft and radiant.

5. Perfumes in Egypt

Egyptian ladies often carried small sachets of scented gum-resins, as is still the custom among the Chinese: indeed, we all know that the dead among the Egyptians were wrapped in aromatics, which have preserved their mummies down to the present day.

6. Greek Perfumes

Greek mythology attributed the invention and use of perfumes to the immortals, and, according to the Fable, men only learned of them through the indiscretion of Œnone, one of Venus' nymphs.

When the gods of Olympus favored a mortal with their visit, they left behind them a scent of ambrosia, an unequivocal sign of their divine nature.

The custom of anointing the bodies of the dead with perfumes was not peculiar to the Jews; all the peoples of antiquity seem to have practiced the same ceremonial in this respect; thus we find in Homer that Venus herself watched over the remains of Hector night and day, pouring over him a precious balm.

The Greeks, moreover, were very fond of perfumes and had made remarkable progress in the art of the perfumer; in this respect, they went so far as to enclose their clothes in scented chests, as we learn from Homer's account of Ulysses, and, according to Athenaeus, they had cassolettes that spread sweet smells in the air while they were at table. Like the Romans, they had the custom of crowning themselves with roses at feasts, and the Athenians' most esteemed wines were perfumed with violets, roses and various aromatics; that of Byblos, in Phoenicia, was especially remarkable in this respect. The luxury of perfumes was even taken so far that a law passed by the wise Solon forbade Athenians from using them.

Among the Lacedemonians, this luxury was always outlawed, and perfumers were banished from the city as people who lost oil, for the same reason that all those who dyed wool were rejected, because they destroyed its whiteness.

Despite the prohibitions of Solon and Lycurgus, the taste for perfumes soon became widespread in Greece, and was pushed to a degree of refinement that had never been reached before and has never been surpassed since.

Although the Orient supplied the Athenians with the most highly prized gums and essences, they considerably expanded the list of fragrant plants already in use.

Apollonius, a disciple of Herophilus, wrote a treatise on perfumes. The best iris," he says, "comes from Elis and Cyzicus; the best rose essence is made in Pharsales, Naples and Capua; crocus essence is superior in Soli, Cilicia and Rhodes; essence of spikenard from Tanius, vine leaf extract from Cyprus and Adramyttium; marjoram oil and apple extract from Cos; palm oil from *Cyprinus* (Gladiolus) from Egypt; the best afterwards from Cyprus and Phoenicia, especially Sidon. The perfume known as *panathenaecum* is made in Athens; in Egypt, the perfumes known as *meto-pian* and *mendesian* are prepared with superiority; however, the quality of each perfume is due to the substances and operations rather than to the country itself."

The boxes in which the ointments were kept were usually made of alabaster, elegantly adorned, and would form an important item in the jeweller's brief; they were called *alabastra*. They were also *onyx* vases. These preparations were preserved in oil and colored red, with cinnabar or sorrel (Pliny); but, if we are to believe a passage in Alexis' *Colon,* this prodigality itself was far surpassed: "For to perfume himself, he did not dip his fingers in alabaster, the ordinary custom of bygone times, but released four doves fully impregnated with essences, not of a single species. Each carrying a particular perfume different from the others, they hovered over us, and with their damp wings rained their fragrances on our dresses and clothes; I too, don't be too jealous, gentlemen, was doused with essence of violets."

The room in which a meal took place was always perfumed, either by burning incense or by scattering eau de senteur on the furniture - an unnecessary precaution considering the profusion with which the guests themselves covered themselves in essences. Each part of the body had its own particular fragrance: mint was recommended for the arms, palm oil for the cheeks and chest; an ointment made from marjoram was applied to the eyebrows and hair; essence of ground ivy was used for the knees and neck; Quince provided a useful essence for lethargy and dyspepsia; the fragrance of grape leaves kept the mind lucid and that of white violets was favorable to digestion.

As part of the athletic exercises at the Olympic Games, wrestlers and pancratiasts (wrestlers who fought in pancrati) used to oil their limbs to make them more supple. Perfumes from Athens enjoyed a great reputation, as we can see from a curious fragment handed down by Athenaeus. This passage shows us where the most esteemed of each kind was obtained in his time, "a cook from Elis, a cauldron from Argos, wine from Plionte, tapestries from Corinth, fish from Sicyone, cheese from Sicily, perfumes from Athens and needles from Boeotia": just like the Romans, who esteemed above all the roses of Paestum.

In Greece, perfumers' stores were open to all: they served as meeting places, where the interests of the state were discussed, fashions decreed, scandalous stories told, and in Athens it was said: "Let's go to the perfume shop!" - as we say, "Let's go to the café."

The fashion for perfuming the head at banquets is said to have stemmed from the idea that the stimulating effects of wine would be better tolerated with a damp head, just as a patient consumed by a burning fever feels relieved by the application of a wet compress. Aristotle, better guided by his habit of observation, made a different and truer argument, attributing the large number of men with grey hair to the drying nature of the ingredients of which ointments were composed, and he was not alone in this opinion. It is not without intention that Sophocles depicts Venus, the goddess of pleasure, perfumed and looking at herself in a mirror, and that he shows Minerva, the goddess of reason and chastity, softening her limbs with oil before engaging in gymnastic exercises.

Chrysippus sought in the etymology of the word a reason to reject the thing; but this far-fetched argument only served to expose him to the mockery of an ancient joker, who said in this connection that, without physicists, there would be nothing so silly in the world as grammarians.

Socrates proscribed all perfumes: "Slaves and free men," he said, "when perfumed, smell the same. This criticism made little impression on his pupil Aeschines, who became a perfumer, incurred debts and tried to borrow money against the value of his stock. Alexander was more sensitive to the observations of his preceptor Leonidas, who reproached him for lavishing incense in sacrifices: "It will be time," his master told him, "to show yourself as generous when you have conquered the countries that produce incense." The king remembered the lesson, and when he had conquered Arabia, he sent his old preceptor a considerable supply of frankincense and myrrh.

7. Roman Perfumes

From Greece, perfumes quickly found their way to Rome; and although the sale of perfumes was at first strictly forbidden, their use became more extravagant every day. To get an idea of this, all you have to do is read the Latin poets, especially the comic ones.

The Romans, who had conquered Egypt, India and Arabia, obtained huge quantities of perfumes from these regions, to which they added those produced in Italy and Gaul.

Fragrant rush was their most common perfume, reserved exclusively for courtesans; the most esteemed were Paestum roses, spikenard, *megalium*, *telinum*, *malabathrum*, *opobalsamum*, *cinnamomum*, etc. They used them in wild profusion to perfume their baths, bedrooms and beds. Like the Greeks, they used them for different parts of the body, and mixed them with wine.

When they had a stage show, the *velarium* covering the amphitheater was impregnated with eau de senteur, which was released as a fragrant rain over the actors and spectators, and the Roman eagles themselves were scented with the finest essences before battle, a ceremony that was repeated when victory had been theirs.

Let's quote a few facts to illustrate the Roman abuse of perfumes: at the funeral of his wife Poppea, Nero had more incense burned at the stake than Arabia produced in an entire year. In an earlier era, Plancius Plancus, proscribed by the triumvirs, had been betrayed by the essences he wore: the scents escaping from his retreat revealed him to the soldiers sent in pursuit.

Pliny cites a large number of cosmetic preparations used by the Romans, who dyed their hair black with St. John's wort, myrtle, cypress, boiled leek peel and walnut stain.

A mixture of oils, ashes and earthworms prevented them from turning white; myrtle berries prevented baldness, and bear fat, even in those days, made hair grow, just as it does today!

Hair was made blond with vinegar lees, or quince juice mixed with privet juice, which was practiced by courtesans who were forbidden to wear black hair; it even appears that some refined women dyed their hair blue.

It was also customary for Roman women to darken their eyebrows.

Carmine was used to color the cheeks, mandrake to erase facial scars and, in addition to simple substances, the perfumers of Rome composed a host of mixtures that can be seen in Pliny or in Ovid's *Cosmetics*, some of which earned their authors the right to have their names passed down to posterity. Martial has preserved for us those of Niceros, Cosmus, Folia, etc. The taste for perfumes was not, moreover, peculiar to certain peoples: it is found in all, and can be traced down through the ages to the present day.

8. *Perfumes in Italy*

Following an ancient custom, every year the Pope in Rome blesses what is known as the "golden rose". This flower, made of the purest gold and adorned with precious stones, is perfumed with balsam and incense. Her Holiness recites prayers explaining the meaning of the blessing, after which she takes the flower in her left hand and blesses those present. Mass is then celebrated in the Sistine Chapel. These golden roses are usually sent to sovereigns, sometimes to princes; other times, though rarely, to cities and corporations. The one for 1862 was given to the Empress of France, and the one for the previous year to the Queen of Spain.

9. *Perfumes in England*

Thanks to Stow, we know exactly when perfumes were first introduced.

"It was not until the fifteenth year of Elizabeth's reign that the Right Honourable Edward de Vere, Earl of Oxford, on his return from Italy, brought back gloves, sachets, a perfumed skin pourpoint and various other novelties. That same year, the Queen had a pair of perfumed gloves, adorned only with four colored silk bouffettes or roses. Elisabeth was so pleased with this new adornment that she had herself painted with her gloved hand, and for a long time was referred to as "the perfume of the Earl of Oxford".

Never in the three kingdoms were perfumes and cosmetics richer, better prepared, more costly or more delicate than during Elizabeth's reign. Her Majesty had a particularly keen sense of smell, and nothing hurt her more than an unpleasant odor. Perfumes and cosmetics of all kinds were in general use at the time. Cosmetics and other toiletries were kept in boxes impregnated with some favorite scent, known as *"sweet coffers"*. This expression is constantly used by old writers. These boxes were considered to be an essential part of the furniture of any room of honor; their richness of form was a sure sign of the taste and generosity of the master of the house. Flasks of essence used for ordinary toiletry care were called *"custing bottles"*; *pomanders, which were* originally intended only to prevent infection, like camphor sachets today, but soon became a luxury item among people of quality, were balls of scented pasta worn in the pocket or around the neck. Soon they became the occasion for the most charming works of goldsmithery, and were often offered as souvenirs or tokens of satisfaction, as snuffboxes were later made. Queen Elisabeth received several as New Year's gifts, including the items shown in figure 1.

Fig. 1 - A scent ball from Elizabeth's time, reproduced with the permission of the curators of the Science and Art Exhibition, based on one now in the South-Kensington Museum. Each division of the ball has a separate compartment designed to hold different perfumes in powder or paste form.

Perfumed gloves were also fashionable, and Elisabeth wore a coat of perfumed Spanish skin. Even her shoes were scented. The city soon imitated court usage, as we can see from the frequent allusions to it in the dramatic writers of the time.

Indeed, these writers are full of satirical observations on the frequent, excessive use of essences and perfumes at the time.

As a testimony to the spirit of the last half of the 17th century, we can cite here an Act of Parliament of 1770. It states that "any woman of any age, rank, profession or condition, whether a virgin, daughter or widow, who, from the date of the said Act, *deceives, seduces* or *entices* into marriage any of Her Majesty's subjects by means of *perfumes, false hair, Spanish crepons* (a kind of woollen cloth impregnated with carmine and still used today as a rouge under the name of "crepon blush"), steel busks, baskets, heeled shoes and false hips, will incur the penalties established by the law currently in force against witchcraft and other manoeuvres ; and that the marriage will be declared null and void.

10. Perfumes in France

Closer to home, we still find cosmetics in the limelight.

Gregory of Tours tells us of the art with which Clotilde, Brunehaut and Galsuinte enhanced the radiance of their charms; he also tells us that the Franks and Gauls knew several artificial wines, which this author calls vina *odoramentis immixta*. The author of the novel of Persée, Forest, also notes, in describing a festival, that "each and every one had a rose hat on his chief". Mathieu de Coucy recounts that, at a banquet given by Philippe le Bon, Duke of Burgundy, a statue of a child was seen "pissing rose water".

In the early days of the French monarchy, it was customary to place cassolettes and perfumes, exhaled by fire, in open coffins; such cassolettes have been found in the tombs of one of the churches in Paris.

Among the gifts Haroun al-Rachid sent to Charlemagne were perfumes, and the Arab invasion of Spain brought with it previously unknown ointments and cosmetics. The Crusades brought new perfumes to Europe, and the discovery of America introduced us to cocoa, vanilla, Peruvian balsam, Tolu balsam and more.

During the Renaissance, the scepter of perfumery was held by Italian artists brought in by François I and Catherine de Médicis; this era can be compared to that of Martial for the abuse of pastes and ointments, perfumed gloves and all the refinements of the art. Historians attest that Diane de Poitiers, thanks to the cosmetics she used, retained all her charms until an age when her rivals had given up trying to please; it is claimed that she got her secrets from Paracelsus. Alongside the châtelaine d'Anet shone the Marguerite des marguerites and the heroines celebrated by Brantôme, who demanded all the resources of Italian cosmetics. It was at this time that the works of Saigini, Guet, Dettazy, Isabella Cortese and Marinello on cosmetics were published, all of which dealt with this art in a remarkable way. Under the Valois, the use of perfumes went as far as abuse; pastes, ointments and the mask of Poppea, found for Henri III and his sweethearts, led to the kind of reaction against perfumes and cosmetics that occurred during the following reign; but the practices of René the Florentine, the gloves of the Queen of Navarre and those of the beautiful Gabrielle contributed to this repulsion, just as the sellers of powder later frightened the court of Louis XIV.

After being neglected under Henri IV, who used scents and ointments sparingly, perfumes regained favor at the court of Louis XIII, under the influence of the beautiful Anne of Austria.

Almond paste and cocoa and vanilla creams, imported from Spain, were used to whiten the hands and shoulders of the beautiful ladies of the court and the Hôtel de Rambouillet; it was at this time that the most precious and sought-after names, borrowed for the most part from the vocabulary of Le Tendre, were used to designate cosmetics; they were outlawed a second time by Louis XIV, who hated them, and they were definitively revived under the Regency. The almost secular beauty of Ninon de Lenclos shows the progress that the art of perfumery had made by then.

With the Regency, perfumes returned to the court; it was at this time that the maréchale powder was invented. Powders, blushes and ointments were also used. Ninon de Lenclos kept her beauty until she was sixty, and Cagliostro later sold a marvellous recipe to La Du Barry that kept her young and beautiful until the limits of old age. In his final years, the Maréchal de Richelieu lived in a fragrant atmosphere, with bellows blowing through his apartments.

With Marie-Antoinette, the taste for perfumes became purer; instead of strong, lively scents, violet and rose were preferred. The same preferences have been retained to this day.

A perfume in general use, even today, was invented by a member of Rome's most ancient nobility, called Frangipani, and still bears his name; it's a powder composed of all known aromatics, in equal proportions, to which iris root is added in equal weight to the whole, with 1% musk and civet. A liqueur of the same name, invented by his grandson Mercurio Frangipani, also enjoys general favor: it is prepared by infusing frangipane powder in rectified wine spirit, which dissolves the fragrant principles: this perfume has the merit of being the longest-lasting of all.

Here is the origin of frangipane:

In Rome, there is a family that bears the name Frangipani; this name is said to derive from a certain function that one of its authors fulfilled in the church, that of presenting the consecrated bread in a particular ceremony: *frangipani* literally means "broken bread", and comes from *frangere panem*, "to break bread"; hence we have frangipane tarts which - as good housewives know - are all made of crumbled bread. One of the members of this ancient family, Mutio Frangipani, served in France in the Pope's army during the reign of Charles IX; it was this gentleman's grandson, the Marquis de Frangipani, Marshal of the armies of Louis XIII, who invented a kind of perfumed glove that came to be known as the "frangipane glove".

Now that we've talked about perfumed gloves, which, incidentally, have long ceased to be in use, we'd like to share a few curious details about the glove and perfume trade.

Before the French Revolution, the perfume industry was governed by the guild system. In 1190, Philip Augustus granted the perfumers statutes which were confirmed by King John on December 20, 1357, and by royal letter from Henry III on July 27, 1582, and which governed the industry until 1636. Under Colbert, who gave great impetus to French industry, the perfumers or parfumeurs-gantiers, as they were called, obtained patents registered with the Parliament, proving their acquired importance;

their brotherhood was established at the chapel of Sainte-Anne in the Church of the Innocents; by patents granted on July 20, 1426 by Henry II, King of England, who called himself "King of France" during the troubles that marked the reign of Charles VII, their arms registered in the general armorial in France are "argent à trois gants de gueules, au chef d'azur chargé d'une cassolette antique d'or".

The Revolution made itself felt in the perfume industry; every perfume had a strange name; there were guillotine clothes, Sanson's ointment, and so on. It was from this period that the perfumery industry was transformed by the support of science; it was under the Directoire that the belles dames revived the perfumed baths of Rome and Greece. Mme Tallien, emerging from a strawberry and raspberry bath, was gently rubbed with sponges soaked in milk and perfume.

Emperor Napoleon I was very sensitive to the action of perfumes; he himself poured eau de Cologne over his head and shoulders every morning; Empress Josephine had the taste of a Creole for flowers and perfumes: she had brought cosmetics from Martinique, the use of which she never abandoned. It was at this time that the consumption of perfumes was at its greatest.

The appearance of *eau de Cologne* marked a milestone in the history of perfumery.

We thought it would be interesting to explain the circumstances in which it made its entry and gained considerable importance in the industry.

Eau de Cologne was invented by Jean-Paul Feminis, who lived in the town of Cologne, from which it takes its name, in the mid-17th century. In 1806, one of his descendants left Cologne to found the Jean-Marie Farina company in Paris. It is to him that the company owes its worldwide renown. Messrs Roger and Gallet have owned the company since 1862.

Jean-Marie Farina eau de Cologne is as pleasant as it is healthy. Its hygienic properties have led doctors to include it in many Codex formulas.

Eau de Cologne strengthens and refreshes the skin, restoring its suppleness. It dissolves tartar from teeth without attacking enamel. Used in a bath, it restores elasticity to muscles and tones the body; it soothes headaches; vaporized, it corrects stale air in apartments and purifies them.

Scented gloves

As mentioned above, the Paris *glove-makers and perfumers* constituted a considerable guild.

As glove makers, they had the right to sell gloves and mittens of all kinds of materials, as well as the skins used for gloves, and, as perfumers, they had the privilege of perfuming gloves and selling all kinds of perfumes. Perfumed skins were imported from Spain and Italy to make gloves, purses and gibecières. These skins were very expensive and very fashionable, but their penetrating scent made them no longer used for gloves; however, Spanish skins are still much sought-after for scenting writing paper. As for gloves, Savary adds: "Quantities of perfumed Italian and Spanish gloves used to be produced; but their strong musk, amber and civet odors, which could not be sustained without discomfort, meant that their fashion and use were almost lost; the most esteemed of these gloves were franchipane and neroli gloves".

There are a large number of recipes for perfuming gloves, some of which are of particular interest. Here is a book entitled *Secreti della signora Isabella Cortese ne'quali, si contengono cose Minerali, Medicinali, Artificiose, ed Alchimiche e molte dell' arte perfumatoria appartenenti a ogni gran signoria* (Venezia, 1574). In it, we find instructions for the superior preparation of gloves with musk and amber, as well as "an excellent preparation of gloves without musk".

In French cuisine, the word "franchipane" or "frangipane" is used to designate a type of pastry made from almonds, cream and sugar. In the West Indies, it is used to designate the plant and fruits of *Plumeria alba* and *Plumeria rubra* L., because, according to Mérat and de Lens, "we find in these ripe fruits the taste of our franchipanes"; if these fruits are really edible, it is remarkable that neither Sloane nor Lunan mention the fact. In any case, the French name for *Plumeria* is "frangipanier".

II. SMELL AND ODORS

As an art form, perfumery will not rise to the heights it needs to reach as long as those who trade in it carefully conceal their processes. No industry that operates under a cloak of secrecy can grow or gain general prominence. I would willingly agree with the Greek physicians who, every year, inscribed in the temple of Aesculapius all the cures they had performed and the means they had used to achieve them.

As for the mystery that surrounds industry," says Professor Solly, "I'll limit myself to saying that, in my opinion, manufacturers would be much better off if they were more willing to benefit from the experience of others, and at the same time less suspicious and jealous of the so-called secrets of their trade. It is a great mistake to believe that a skilled manufacturer is one who has carefully guarded the secrets of his manufacture, or that particular ways of making certain objects, processes unknown in other factories, mysteries beyond the intelligence of the vulgar, matter in any way to the prosperity of a factory or the success of a business.

"In times of ignorance, everything was a secret, a mystery or a spell in the hands of associations, guilds or communities. In those days, those who knew lived at the expense of those who didn't, and no one sought to acquire knowledge except as a means of gaining the upper hand over their neighbors. Science, thus separated from reason and, as it were, stripped of its innocence, was naturally treated as a kind of witchcraft, and anyone who was one step ahead of the intelligence of his contemporaries was often burned as a practitioner of black magic. As we know, most of those who suffered this cruel treatment had only themselves to blame for the opinion that became fatal to them, for having deliberately misrepresented their knowledge. There are still secrets today, and many are prized as highly and guarded as carefully as the secrets of art in the Middle Ages. But an atmosphere of mystery is rarely conducive to the development of public prosperity or even private advantage. Early manufacturers had no secrets. They were ready to open their workshops to all foreign visitors; and even when they had found a way of saving labor or materials that was unknown to their colleagues, they did not keep it to themselves. They have more confidence in the spirit of progress, in an energy that is always ahead of its time, than in the exclusive possession of this or that process. Small minds don't understand this. They're always looking for the process that will open the door to fortune and show them the way to opulence" (Solly).

Smells and Odors

It should be added, however, that these remarks apply more to the industries that produce fragrant raw materials than to perfumery itself.

There was a time when "distilled waters" and "cordials" were prepared in the *still room* and administered to the manor's guests and servants as a specific treatment against illness. However, the *still room maid* still retains her name, although she is hardly ever called upon to perform her former duties.

1. *The Sense of Smell*

Of the five senses, the sense of smell has received the least attention. But as science progresses, the various faculties with which the Creator's wisdom has been pleased to endow man develop more and more, and the sense of smell will receive its share of education like those of sight, hearing, touch and taste.

The influence of this sense on the constitution is quite remarkable. One smell immediately causes disgust, nausea and vomiting; another gives the mind a feeling of cheerfulness and well-being. Such, for example, are the fragrances one breathes in the countryside on a spring morning, or the gentle sea breezes impregnated with the fortifying emanations that escape from the grasses washed up on the shore. The first time an inland inhabitant breathes sea air, it has an extraordinary effect on his entire nervous system.

2. Smell Theory

The smell of fields as hay is being mown, the scent of a garden as the day draws to a close, charm and recreate the spirit.

Odors can spread so far that it's hard to believe that the existence of an odor always and necessarily implies the existence of a body. It often seems that a smell acts as an imponderable agent rather than as a physical substance. It is obvious that various substances produce certain odors, but it is not equally certain that these substances are themselves odors. Ultimately, the best way, in my opinion, to understand the theory of odors is to consider them as particular vibrations that affect the nervous system, just as colors affect the eye, just as sounds affect the ear.

The vibrations may well be due to the chemical actions that essences and perfumes undergo on contact with oxygen in the air; in fact, they can all be made odorless by volatilizing them away from contact with oxygen. Essences thus deprived of odour instantly regain it on contact with air. Every chemical combination produces vibrations that give rise to luminous and electrical phenomena; in some cases, other vibrations are produced that can affect the olfactory nervous system: for each odor, the speed of the vibrations would be different.

The analogy between color and sound has long been recognized. The ancients were well aware of this relationship when they turned the musical scale into a *chromatic* scale. Bacon and, since him, a host of writers have dealt with this subject, and many have tried to demonstrate that the harmony of colors is in harmony with the melody of the scale.

G. B. Allen, in studies on the analogy between colors and the musical scale, establishes that all composers of merit feel this analogy and that all their works bear witness to it.

This is how Field arranges the ladder:

Blue	Purple	Red	Orange	Yellow	Green	Green
C	*D*	*E*	*F*	*G*	*A*	*B*

And here's how he proves the analogy. These three primitive colors, blue, red and yellow, combined or opposed, produce the most perfect harmony, as do the sounds *do, mi, sol.* The metrochrome and the monochord also prove the accuracy of this double tuning. The first of these two instruments shows us that, in pure white, there are eight shades of blue, five of red and three of yellow, and so on. The second, that eight parts of a string will give the sound *C*, five the sound *E* and three the sound *G*. This curious agreement certainly proves the existence of a universal law of harmony.

To measure the intensity of light and sound, we have a method based on the speed with which both travel through space.

Struck, along with several other observers, by the close analogy that exists between the forces that affect our various senses, and particularly those that affect the organs of smell and hearing, but unable as yet to find any adopted type by means of which I could in some way measure the intensity of an odor as one measures that of a sound, I undertook a series of experiments in order to discover one.

I had long noticed that when alcoholic solutions of various essences mixed together were allowed to evaporate in the open air, they underwent a kind of natural analysis, i.e. the more volatile ones evaporated first, while the less volatile only disappeared later.

Seeing the same thing happening over and over again when the essences were the same, I soon recognized that a kind of definite, inherent power abandoned each odorous body or remained faithful to them for a more or less long time. It's this power that I call the "speed of smell", or, in other words, the "power of volatility". Now, I find a connection between this power of volatility and the way an odorous substance affects the sense of smell. I don't mean to say that, because a body has a high power of volatility, it will affect the organs of smell in one way, and that a body with a lower power of volatility will affect them in another way. I know that there are volatile bodies, such as mercury and water, that have no odor, a phenomenon due mainly to the fact that these vapors are insoluble in the secretions that lubricate the nasal membranes. What I have observed, however, is that substances which are exhaled naturally or extracted by man from plants or animals, and which are recognized as odoriferous bodies, affect the olfactory

Smells and Odors

nerves as a direct result of their power of volatility, a power which I call the "speed of odor" because it has an action on the odor of a body as long as that body is soluble in the pituitary secretion.

The power of volatility or speed of odor should not, if we wish to be consistent with what precedes and follows, be defined and explained as it is in this paragraph, and especially we should not say that the odors produced are as a direct result of the solubility of the vapors in the liquid from the pituitary secretion. For certainly water vapour is soluble in this secretion and is odourless; rather, it could be said that the power of volatility of the essences, or the rapidity with which they evaporate, would always be related to the speed of vibration produced or the rapidity with which the odorous waves would propagate ; if this speed were not high enough, there would be no odor perceived, in the same way that the ear cannot hear sounds that do not correspond to at least 60 vibrations per second; the liquid that lubricates the olfactory membrane, necessary to perceive odors, would have the role of increasing the sensitivity of the nerves, which would thus be more capable of perceiving odors.

Thus, bodies with a very low degree of volatility are those known as "strong odors"; those, on the contrary, with a high degree of volatility are weak and delicate odors. Here we see the analogies of certain effects on the senses. The slowest-propagating sound waves produce the loudest sounds; the slowest-propagating odour waves produce the strongest smells.

In speaking even briefly of the physiological action of odors, it is necessary to remind the reader of the distinction between substances that irritate the nerves of tactile sensitivity, and those that communicate the impression of an odor to the olfactory nerves, because certain pulverized solids, such as glass dust, soap powder, tobacco and certain gases like chlorine, ammonia, etc., excite the pituitary membrane. The effects produced by these substances are those of a body *touched*, not *smelled*.

In other words, the more or less irritating local mechanical action of certain bodies on the pituitary membrane should not be confused with the action of odors themselves on the olfactory nerves.

There are many people who are *anosmic*, i.e. deprived of the sense of smell, and for whom all smells are the same. "They have noses, but do not smell. They resemble people who are deaf, because they cannot hear, and others who are blind, because they cannot see. Anosmia is more frequently the result of a lack of exercise of the sense of smell than of a natural predisposition. On the other hand, some people are *hyperosmic*, i.e. have an excessively subtle sense of smell.

When planting a garden intended as much to charm the sense of smell as to delight the eye, one could do no better than be guided by the data we're about to set out in choosing the flowers to cultivate. Similarly, those who admire the heavenly fragrance of a garden at dusk could not neglect the cultivation of nocturnal flowers without losing all the voluptuous emanations they exhale into the atmosphere, emanations all the more beyond the reach of the philosopher-chemist as they are subtle and ethereal.

3. Odor Harmony

Smells seem to affect the olfactory nerve to certain degrees, just as sounds affect the auditory nerves. There is, so to speak, an octave of odors, like an octave of notes; certain fragrances blend together like the sounds of an instrument. Thus, almond, heliotrope, vanilla and clematis go very well together, each producing more or less the same impression to a different degree. On the other hand, we have lemon, lime, orange peel and verbena, which form a higher octave of odors, and combine similarly. The analogy is completed by what we call "half-odors", such as rose with rose geranium as a half-tone; petit grain, neroli followed by orange blossom. Then come patchouly, sandalwood and vetiver, and several others that fit into each other.

Using the flowers we already know, we can obtain, by mixing them in certain proportions, the fragrance of almost any flower, except jasmine.

As Dickens puts it: "Is the jasmine the mystical Meru, the center, the Delphi, the Omphale of the flower world? Is it the starting point of all fragrance, the indivisible, elusive unity? Is jasmine the Isis of flowers, with veiled head and hidden feet, who is loved by all and revealed to none? Charming jasmine! If this is so, then the rose must come down from its throne and surrender its queenly crown to unparalleled beauty. Revolutions and abdications are moving games. What if we were to spark a civil war in the gardens and proclaim Jasmine emperor and king of the flowerbeds!"

The fragrances of some flowers so closely resemble those of others that one is almost tempted to believe they are identical; at least, if they are not identical at the moment they are exhaled from the plant, they seem to become so through the action of the air. We know that some of them are really identical in their composition, even though they come from completely different plants. From this identity we can conclude that, sooner or later, chemistry will produce one with the other, since for many it's only a molecule of water or an atom of oxygen that makes the difference. It would be a great thing to extract a finer, more expensive essence from a less valuable, yet sufficiently widespread product.

A very small quantity of almond essential oil in a bottle containing plenty of oxygen changes into another fragrant substance, benzoic acid, which can be seen forming crystals on the dry walls of the bottle. This metamorphosis is a natural demonstration of the above theory.

The artificial formation of some essences and their transformation into each other have already been carried out, as we shall see later; we have respected the English text, and translated it faithfully; but we must point out that pure benzoic acid is perfectly odorless, and if that of the trade often presents a particular odor, it is because it contains foreign matter.

For the "ignorant" nose, all smells are:

Fig.2 - Odor range, bass or G clef.

Fig. 3 - Odor scale, bass or F clef.

But the nose, civilized by pleasure or interest, becomes the most delicate and sagacious of organs. Wine and tea merchants, druggists, tobacco importers and others have to give the olfactory apparatus a real education. A hop merchant dips his nose into a bag, inhales the scent of the flower, and then says what price he wants to give for it.

We need to remember smells, and the tenacity with which they are fixed in the memory is a fact to be noted here: without this memory, the various merchants we have just mentioned would be quite embarrassed. An experienced perfumer sometimes has two hundred scents in his laboratory, and knows how to distinguish each one by name. What musician could, on a keyboard containing two hundred notes, recognize and name the key struck without seeing the instrument?

In the range above, I've tried to place the name of each scent in the position corresponding to its effect on the senses.

I've deliberately chosen the smells that are most commonly used in perfumery, but I'd like to make it clear that all smells, whatever their source, can be classified in the same way. I don't know of a single odor in a chemical laboratory - and there are quite a few - to which I cannot assign its corresponding place.

There are smells that admit neither sharps nor flats, and there are others that would almost make up a whole range on their own, thanks to their various nuances. The class of smells that contains the most varieties is that of lemon.

When a perfumer wants to make a bouquet of primitive scents, he must take scents that go well together; the perfume will then be harmonious. If you take a look at the range, you'll see what harmony and disharmony are when it comes to scents. As a painter melts his colors, so a perfumer must melt aromas.

When making a bouquet of several fragrances, it's important to mix them so that, when brought together, they create a contrast.

Vanilla's counterpart is lemongrass. The following recipes will give you an idea of how to compose according to the laws of harmony:

Bass.
G Pergular (Pergularia edulis)
G Sweet Peas
D Violet - G Chord Bouquet
F Tuberose
G Orange Blossom
B Aurone

Top.
Bass
C Sandalwood
C Geranium
E Acacia - C Chord Bouquet
G Orange Blossom
C Camphor

Top.
Bass
F Musk
C Tuberose - F Chord Bouquet
A Tonka Bean
C Camphor
F Daffodil

To make a bouquet, all primitive scents must be reduced to a certain degree of strength or potency. Thus, the strength of rose spirit is 95 grams of essential rose oil to one gallon (4.5 liters) of alcohol. But the strength of geranium is 250 grams of essential oil to one gallon (4.5 liters) of alcohol. The difference in odor power between the two essences is like 3 to 8. When it comes to electricity, physicists distinguish between intensity and quantity; verbena can be cited as representing the former and vanilla the latter. Camphor is three times as intense as rose.

"There is," says Sir David Brewster, "in sound and light a property too remarkable to be passed over in silence: two bright sounds may come to produce silence, and two bright lights may come to produce darkness."

If two equal and similar strings, or two columns of air in two identical pipes of the same dimensions, produce exactly one hundred vibrations per second, they will each produce equal sound waves, and these waves will join together to form a continuous sound, double that of each of the sounds heard separately. If the two strings or the two columns of air are not in unison, but only approximately, as, for example, when one vibrates a hundred times and the other a hundred and one times in a second, then at the first vibration the two sounds will form a single sound twice as loud as each of them separately; but one will gradually gain on the other, and at the fiftieth vibration it will be half a vibration ahead. At this point, the two sounds *muffle each other*, and there is an interval of complete silence. Then the sound starts again, gradually increasing and becoming very loud at the 100th vibration, at which point the two vibrations come together to produce a sound twice as loud as the one they make in isolation. A new interval of silence returns at the 150th, 250th, 350th vibration, i.e. every second, while a sound twice as loud as each of the two separate sounds is heard at the 200th, 300th, 400th vibration. When the unison is very imperfect, or when there is a great difference between the number of vibrations that the two strings or the two columns of air produce in one second, the successive sounds and the intervals of silence resemble a hum. With a powerful organ, the effect of this experiment is very curious. The repetition of the sounds *ouoù-ouoù-ouoù* represents the double sound and the interval of silence that results from the total extinction of the two separate sounds.

4. Land Odor

All substances which, in ordinary language, we call "earth", exhale a particular and characteristic odor as soon as they are impregnated with water. Anyone walking along a main road during the summer months who has been caught in a downpour may have noticed the delicious smell that fills the air a few minutes after the rain has fallen, and then fades away. When water is poured over chalk, or rather Spanish white, it gives off a very persistent odor, but one that is not perceptible to everyone; iron oxides, manganese oxides and various other mineral substances give off an odor when wet. For the moment, we confine ourselves to stating the fact, without investigating the cause of these phenomena, and to noting that the odors are certainly not due to any substance pre-existing in the water before it comes into contact with the earth, as the same result was observed when the purest distilled water was used for the experiment. Nor should the observation be restricted to the mixture of earth and water, for when hydrochloric acid is poured over zinc oxide, a pleasant odour is produced, a by-product of the combination of acid and zinc oxide.

The odors emitted by soils can be explained by the presence of organic matter, or by odorous gases absorbed by porous soils and displaced by water; as for the odor emitted when zinc oxide is treated with hydrochloric acid, it could be explained in several ways, and may vary according to circumstances.

5. General Air Odor

Independently of bodies whose chemical nature and mode of formation escape us, and of those which have resisted all attempts at synthesis, a series of chemical reactions take place in nature which, by varying the composition of the air, are perceived by a special odor.

Air is composed roughly of a mixture of 79 parts nitrogen and 21 parts oxygen; but alongside these two principles, there are others, in much smaller quantities, which are of particular interest. In addition to oxygen and nitrogen, air contains argon, water vapor, carbonic acid, ammonia, formene, nitric acid, sodium sulfate, sea salt, etc., the mixture of which forms the air to which our olfactory sense is accustomed.

When, for whatever reason, one or more of these substances increases or decreases in relatively significant proportion, we are immediately alerted by our sense of smell.

Following a thunderstorm, the numerous electric sparks and effluvia give rise to a special body known as ozone, which is nothing other than the product of oxygen condensation. At the same time, nitric acid is formed, which combines with ammonia; various products, which are normally stable, enter into combination, and it is to the whole of these transformations, combinations and modifications that we must attribute the special odor we smell at such times.

In the earth's interior, when heat is high and humidity scarce, organic matter decomposes. This leads to putrefaction and, consequently, the formation of complex products and putrid odors, which make an unpleasant impression.

On the seashore, the air is enriched with sea salt, sodium sulfate and other salts that the sea deposits by evaporation, and which, being carried into space by the wind, are thus suspended in the atmosphere; It is to the presence of these compounds, existing in special proportions, that we owe the invigorating properties of sea air, the powerful action it possesses in curing scrofulous affections, and which, so justly esteemed today, make beaches so popular in fine weather.

The increase in the proportion of hydrogen sulfide in the air is the cause of the rotten egg smell that is so noticeable in the vicinity of the waters of Enghien and Eaux-Bonnes.

The special odor of cemeteries is due to the decomposition of matter contained in the human body, giving rise to the production of *phosphorus hydrogen*.

6. Variability of Fragrance Qualities

> Flowers, the passing gift of a too-quick summer,
> Give, fading, but a fleeting pleasure;
> But their juices distilled into a liquid perfume
> The memory of their living brilliance;
> A glass prison enclosing their scents
> Recall the happy day their presence embalmed.
> Vienna today, winter unleashing its rigors,
> If the flower has perished, let's breathe in its essence.
>
> Shakespeare

A. The Influence of Climate

Flowers exhale fragrances in all climates, but those that grow in warmer latitudes give off more abundant scents, while those in colder regions spread the most delicate odors. W. J. Hooker speaks of the delicious fragrance of flowers in the Skardsheidi valley; we know that violets, primerose (*Primula*) and wild thyme are found there in abundance. Mr. Louis Piesse, exploring the wilds of South Australia with Captain Sturt, writes: "The rains have clothed the land with greenery as beautiful as that of the Shropshire meadows in May, and flowers as sweet as the English violet, to which the white anemone resembles in fragrance. The yellow acacia (cassia), in flower, is magnificent and exhales a penetrating scent."

A writer from Upper Canada, Forster Ker, put it this way: "I'm sending you a few sprigs of our Indian *grass*, whose delicious scent you will not fail to notice. You have nothing in England to compare it with, and I'm surprised your perfumers don't use it. It's very abundant here.

"All countries, all climates offer the Most High the perfumes their soil produces. The most delicious scents perfume the majestic peaks of the Alps; the icy zone is rich in rare perfumes; the Ocean, that wrinkled-faced, grey-bearded old chatterbox, lavishes ambergris on its shores; the scorching regions of the torrid zone intoxicate our senses with the concentrated mixture of their volatile emanations, the delicious aroma of their various products, elusive to chemical analysis."

Although many perfumes come from the East Indies, Ceylon, Mexico and Peru, southern Europe is the only garden truly useful to the perfumer. Grasse (Grasse is located at an altitude of 250 meters as you climb from the sea to Mount Esterel) is the main headquarters of this industry. Within a relatively small circle, the grower finds the various climates best suited to producing the perfect plants for his trade. Not far from the sea, cassia grows without fear of frost, which could destroy an entire crop in a single night; while closer to the Esterel mountains, at the foot of the Alps, violets are sweeter than if they had grown in the warmer exposures where orange and tuberose blossom perfectly. England produces lavender and peppermint. The essential oils extracted from these plants, grown at Mitcham in Surrey and Hitchin in Hertford, fetch a fairly high market price. Grasse produces all rose, tuberose, cassia, jasmine, orange blossom and violet products. Mint of great finesse is also harvested here, as is the highly prized geranium. Sicily gives us lemon and orange, Italy iris and bergamot.

The scent of plants does not all reside in the same parts. For some, it's in the root, as in iris and vetiver (rhizomes or underground stems); for others, in the wood, as in cedar and sandalwood; it's the leaf in mint, patchouly and thyme; the flower in rose and violet; the seed in tonka bean; the fruit in caraway; the bark in cinnamon.

Some plants have several distinct and characteristic scents. The orange tree, for example, yields three: *petit grain* is extracted from the leaves and berries, neroli from the flowers, and an essential oil called *Portugal* from the rind of the sweet orange fruit. For this reason, this tree is perhaps the most precious of all for the perfume maker.

The orange tree that gives the best neroli is the bigarade, not the sweet orange; the essence extracted from its fruit is the essence of bigarade, not the essence of Portugal produced by the sweet orange.

The fragrance of flowers is due, in most cases, to a highly volatile oil contained in small vessels or inner cells, or produced at different times in their life, as when they are in bloom. Some plants produce fragrant gums (resins) by incision, such as benzoin, frankincense, myrrh, etc.; others produce balsams by the same process, which appear to be a mixture of a fragrant oil and an odorless gum (resin). Some of these balms are obtained in the country where the plant is indigenous, by boiling it in water for some time. This infusion is then filtered and boiled a second time, or subjected to evaporation, until the residue has acquired the consistency of molasses. This is how Peru balsam is extracted from Myroxylon peruiferum and Tolu balsam from *Myroxylon toluiferum.* Although these scents are pleasant, they are rarely used in perfumery for handkerchiefs, but some manufacturers use them in soap. In England, they are valued more for their medicinal properties than for their fragrance.

B. Influence of Time of Day and Night

Flowers generally exhale their odors while the sun is shining, or at least during the day; but there are some that smell nothing during the day, and are fragrant in the evening: such are *Cestrum nocturnum, Lychnis vespertina* and some varieties of *Catasetum* and *Cymbidium.*

A small number of flowers owe their specific name, *tristis*, meaning "sad", to the fact that they only smell at night; these include *Hesperis tristis* and *Nyctanthes arbor tristis.*

Recluz, speaking of the effects of solar rays on the flowers of *Cacalia septentrionalis*, says: "I had occasion to observe in 1815, in the King's garden, that the flowers of *Cacalia septentrionalis*, exposed to the action of solar rays, exhaled an aromatic odor, which could be rendered null by intercepting the solar rays by means of a hat or the hand, and then returning them to the contact of solar light."

Morren says that the flowers of *Habenaria bifolia*, which grows around Liège, are completely odorless during the day, while in the evening, usually around eleven o'clock, they give off a very pleasant and penetrating scent. He noted that the fragrance began to be felt at dusk, increased in intensity with the shadows of the night, and faded at daybreak. Two bulbs of this Orchis were placed in two cylindrical vases filled with water in which the plants were completely immersed; one vase was placed in the sun and the other in the shade. When evening came, a delicious odor was felt and continued to be exhaled during the night, but disappeared at sunrise. These experiments led Morren to conclude that the smell of flowers depends on some physiological cause and not on the evaporation of particles, nor on their accumulation in parts of the plants where they originate. He found that aromatic orchids, such as *Marillaria aromatica*, lost their fragrance half an hour after the artificial application of pollen, and that unfertilized flowers retained their scent longer.

Trinchinetti, who also experimented with plant odors, divides fragrant flowers into two classes:

1) Those in which the intermittence of the fragrance is linked to the flower's blooming and sleeping periods.

This class itself contains two subdivisions:

a. Flowers which, remaining closed and odorless during the day, open and exhale a fragrance at night: Mirabilis jalapa, M. dichotoma, M. longiflora, Datura ceratocaula, Nyctanthes arbor tristis, Cereus grandiflorus, C. nycticalus, C. serpentinus, Mesembryanthemum noctiflorum, and some species of silene.

b. Flowers that remain closed and odorless at night, but open and give off a fragrance during the day: *Convolvulus arvensis, Cucurbita pepo, Nymphaea alba* and *Nymphaea caerulea.*

2) Flowers that are always open, but are alternately fragrant and odorless. This class comprises two sections:

a. Flowers that are always open and only smell during the day: *Cestrum diurnum, Caronilla glauca* and *Cacalia septentrionalis.*

b. Ever-open flowers that smell only at night: Pelargonium triste, Cestrum nocturnum, Hesperis tristis and Gladiolus tristis.

The odors emitted by nocturnal flowers are sometimes very intermittent. The flowers of the *Cereus grandiflorus, for* example, are fragrant only at intervals, giving off whiffs every half-hour from eight o'clock until midnight. According to Morren, in one case, the flowers began to open at six o'clock in the evening, when the first odor was perceptible in the greenhouse; a quarter of an hour later, following a rapid movement of the calyx, the first puff was felt; at six twenty-three minutes, a new and very powerful emanation; at six thirty-five, the flowers were all wide open; at a quarter to seven, the odor of the calyx became stronger, albeit modified by that of the petals. The emanations then resumed their usual intervals.

Those who admire the scents of a flowerbed at sunset cannot neglect the cultivation of nocturnal flowers, without depriving themselves of the thousand pleasures of breathing in the aromas they release into the atmosphere, aromas so subtle, so ethereal that they often escape the chemist's analysis.

III. FRAGRANCE AND COSMETICS HYGIENE

The use of cosmetics dates back to ancient times. There used to be a distinction, no longer made today, between *cosmetics* and *commotics*. Anything related to hygiene and intended to beautify the human body was called ars *ornatrix* or *cosmetics*, while anything used to correct natural imperfections or repair the ravages of time was called *ars fucatrix* or *commotique*.

Today, in the etymological sense of the word, "cosmetic" means any substance intended to maintain the beauty of the human body.

Are cosmetics necessary? Are they harmful? To solve such a problem, we need to go into some detail, and deal specifically with each group of cosmetics.

In the first group, we include cosmetics that contain no toxic substances, and whose daily and excessive use is without any inconvenience; but we'll soon see that, even among these, hygiene advises making a choice, and that in certain circumstances, and according to the purpose for which they are intended, one cosmetic should be preferred to another.

The group of cosmetics we'll call "innocent" includes preparations that are sometimes subject to fraudulent alteration. Most of the time, the fraud only concerns the quality of the substances used, and is not likely to affect consumer health. It most often results from the inferior quality of the raw materials used, from the replacement of fats or fine oils by more common fats, or from the use of common essences instead of fine essences in the preparation of alcohols. This does not even constitute fraud, but rather inferior manufacture, and in all cases without danger to public health.

A second group includes cosmetics based on toxic materials whose use, even if restricted, can cause serious injury or illness.

Well-intentioned regulations should make it an absolute rule to conspicuously point out the existence, in cosmetics, of toxic substances whose use, even temporary, can present real dangers, as well as materials which, without being dangerous in themselves, can nevertheless, through daily and immoderate use, be harmful to health. It would certainly be prudent never to use as cosmetics ointments, powders or liquids prepared with lime, lead, copper, mercury, silver salts, arsenic, etc.; the use of these preparations has become so commonplace that it would be quite impossible to forbid their sale without seriously infringing commercial freedom. We should therefore limit ourselves to obliging the manufacturer or retailer to indicate the danger of these preparations on the label.

The manufacture and sale of cosmetics containing poisons raises far more serious, and as yet unresolved, questions, as these preparations are often said to have highly effective therapeutic properties. We will have to examine further whether it would not be appropriate to treat certain cosmetics as medicines, and to require that those containing active ingredients be dispensed exclusively by pharmacists; or whether the administrative authority responsible for public health can take legal action against those selling harmful products, particularly cosmetics. It is essential to explain the composition of these cosmetics, and to demonstrate the dangers to which we are exposed by their daily use, if we are to understand the serious consequences they entail.

The scents and perfumes used in various eras far from our time were used as nature provided them, and the preparations they underwent were limited to little more than burning or powdering them, or applying appropriate solvents to them.

The Greeks, Egyptians, Orientals and above all the Indians burned perfumes: they gave pride of place to powders and perfumed sachets; eaux de senteur, alcoholates and aromatic vinegars were used more particularly by the Romans. The perfumes included in ars *ornatrix* contained no toxic substances: these were roses from Paestum, Phaselia or Campania, iris, narcissus, marjoram, fragrant rush (*Schoenus*), *Schoenanthe odorata, malabatrum, telinum, opobalsamum, carpobalsamum, nards, cinnamomum* (not cinnamon; this was called *cassia*), which came mainly from India. We don't know the composition of these different perfumes, the preparation of which was kept secret by those who traded in them; the raw materials came from Assyria, Egypt and Arabia: this last country was divided into five regions, one of which was the "land of aromatics". The names of some of these perfumers are still famous. Niceros gave his name to *Nicerotian;* there was also a perfumer by the name of Cosmus. Folia, Canidie's companion on Mount Esquilino, gave her name to *foliatum*, a variety of Persian nard, which she prepared using a special process.

In *ars fucatrix* or commotic, alongside the most innocent substances, we find toxic materials such as ceruse, which was used in blushes and to erase wrinkles.

Among odors and perfumes, some are of the nature of resins and essential oils, and can be easily isolated from the products that contain them; others, more fleeting, more difficult to separate, are less well known in their pure state; neither exerts any untoward action, apart from in particular cases and special idiosyncrasies. However, strong odors, breathed in large quantities, can cause fairly serious nervous accidents, intense cephalalgia, sometimes followed by dizziness; it is therefore prudent to breathe perfumes only in small quantities. There are even people who cannot tolerate them in any dose. The famous painters Grétry and Vincent were said to have been very bothered by the smell of a rose. We know of a lady in whom orange blossom caused violent nervous spasms. Ledelius tells of a merchant who suffered from ophthalmia due to the smell of roses. Valmont de Bomare says that the subtle, fragrant parts of the betony flower are so strong that gardeners who pull it out become drunk and stagger as if they had drunk wine. The emanations of mancenillium, *Hippomane mancenilla* (Euphorbiaceae), walnut, hemp, flowering elderberry, etc., are considered dangerous. It is highly probable that these effects are due to the fine particles that escape from these plants when they are ground into powder.

Plant emanations should not be regarded as absolute poisons, i.e. as capable of producing poisoning in all possible circumstances, but only as relative poisons whose effects depend on a lesser nervous susceptibility, or on idiosyncrasy. Historians who claim that gloves, boxes, etc., were once poisoned should not be given much credence. Tales of the poisoning of Henri IV, of a Savoy princess, Jeanne d'Albret, of Pope Clement VII and of a number of other figures who, it is said, fell over backwards after sniffing perfumed boxes and gloves, must be regarded as fabulous. The poisons known to the Ancients were no more active than those we know today, and among the highly odoriferous substances, hydrocyanic acid and bitter almond essence excepted, there is none that can cause death when inhaled. These two volatile poisons (prussic or hydrocyanic acid and bitter almond oil) are released, it's true, in small quantities, when the leaves and flowers of certain Rosaceae plants of the Drupaceae tribe (cherry laurel, peach, bitter almond, etc.) are crumpled, but these

Fragrance and Cosmetics Hygiene

leaves and flowers are practically odorless when intact, and in no case can they produce enough toxic substance to cause serious accidents by simple inhalation.

If natural scents and perfumes, if essential oils extracted from plants, are not capable of causing these accidents, is the same true of certain highly volatile, very subtle artificial essences that chemistry has succeeded in obtaining in recent times, some of which have been introduced into the perfumer's art? We don't think so; most of these products have no particular chemical characteristics, and there is no reason why a synthetic product should be more harmful to health than a substance of plant origin.

Absurd prejudices prevail on this subject, against which we cannot protest enough.

The emanations of fragrant flowers should not be confused with the essences that can be extracted from them. Essential oils can produce no effects other than those inherent to their nature; at most, some of them, when accumulated in large quantities and in thin layers in a confined area, may vitiate the air by resinating through oxidation, and producing carbonic acid; but these are exceptional cases that will never be determined by perfumed cosmetics.

The same cannot be said for fragrant flowers accumulated in a confined space; it is always unwise to stay in apartments where fragrant flowers are present, including tuberose, jasmine, magnolia, etc., because the phenomenon here is complex. In fact, in addition to giving off odors of varying strength, these flowers vitiate the air with carbonic acid, which may be laden with a little carbon monoxide, a deleterious agent that Mr. Boussingault has described as one of the products of plant respiration in certain circumstances. Every plant, in addition to its odor, is a hotbed of more or less fearsome exhalations.

A well-known experiment consists in placing a rose, stripped of its leaves, under a hermetically sealed glass bell in the evening. During the night, the rose absorbs oxygen from the air inside the bell, and in exchange gives off carbonic acid; the next day, if a lighted candle is placed near the rose, it goes out. A flower forgotten in a bedroom can cause headaches, nausea and dizziness. However, no jar of ointment, whatever its aroma, has ever been accused of similar misdeeds.

Apart from a few exceptional cases, there's no danger in breathing in small quantities of scents and perfumes; but the abuse of perfumes throws the mind and body into a kind of languor. These enervating characteristics are especially characteristic of fine or somewhat bland scents, such as rose, lily, jasmine and tuberose. Let's recall with Constantin James those Asians who, to numb the woman in the slavery of the harem, surround her with an atmosphere impregnated with fragrant effluvia, that perfumed court of Louis XV which was, of all courts, an effeminate one.

As for aromatic, penetrating scents, such as those derived from lavender, thyme, mint and verbena, they revive and restore; one degree more, and they can become an effective brain stimulant. Breathing acetic acid ("English salt") or ammonia will suffice to prevent or dispel a fainting spell. (Constantin James.)

1. Hair Hygiene, Depilatory Preparations

Few customs have more ancient origins than that of painting the face, dying the hair and darkening the eyelashes and eyebrows to enhance beauty. In Egypt, it is a general custom among women of the upper and middle classes, and very common among those of the lower classes, to blacken the edges of the upper and lower eyelids with a powder they call "kohol". The kohol is applied with a small wooden stylus, thinned and blunted at the tip. Soak it occasionally in rosewater, then dip it in the powder and move it along the edge of the eyelids. This operation is thought to give a very gentle expression to the eyes, making them appear larger. This is probably what Jeremiah is referring to when he says: "Though you split your face (eyes) with color, you will make yourself beautiful in vain" (Jeremiah, IV, 30).

A peculiar custom of Moorish and Arab women is to draw bunches of bluish dots or other small figures between the eyes, to which they apply a color that makes them indelible. The chin is also tattooed in the same way: a small blue line running from the tip down to the throat. The eyelashes, eyebrows and the edges and tips of the eyelids are also colored black. The soles of the feet and sometimes other parts of the foot, down to the ankle, the palms of the hands and the nails are dyed a yellowish red with the leaves of a plant called "henna" (the *Song of* Solomon mentions this plant under the name of *camphire*), "henna" or "alkanna" from Cyprus and Egypt, *Lawsonia inermis* (Salicariae). Its leaves, which are dried for this purpose, are somewhat similar to myrtle. They are crushed and made into a paste with limewater, which is applied to the skin, hair and nails and left for several hours; the color thus imprinted remains for weeks. The tops of the hands are also often painted in this way, and decorated with various designs. On feast days, the cheeks are painted brick-red; a small red line also marks the contour of the temples.

Persians, young and old, dye their hair and beard every eight days. We'll have a chance to examine two powders they use for this purpose; Feroukh-Kan gave them to Professor Trousseau: one dyes the hair golden yellow: it's henna; the other dyes it blue: it's most certainly an indigoferous plant whose name is unknown to us. First apply the henna, making a paste with water, cover the head with it and, after half an hour's contact, apply the blue powder in the same way, obtaining a magnificent "raven-wing" black color.

Similar practices still exist in Persia.

In her book *Glimpses of Life in Persia*, Lady Sheil says of the Shah's mother: "The palms of her hands and the tips of her fingers were dyed red with a herb called 'henna', and the edge of the lower eyelid was colored with antimony. All Kadjars naturally have large, arched eyebrows; but women are not content with what nature has given them, they

Fragrance and Cosmetics Hygiene

enlarge them and double their actual proportions by extending them with large lines drawn with antimony. Their cheeks are covered in blush, as is the invariable custom of Persian women of all classes.

"In Greece, to dye eyelashes and eyelids, essence or gum labdanum, *Cistus creticus* (Cistinea), is thrown onto embers; the smoke is intercepted with a plate to collect the black. Here's how I've seen this preparation used: a young girl, seated on a sofa, legs crossed as usual, closing one of her eyes, takes the two eyelashes between the thumb and forefinger of her left hand, pulls them forward, then introduces through the outer corner a kind of pin or stylus previously dipped in the black smoke. When the stylus was withdrawn, the particles of color adhering to it would stop between the eyelashes and remain there."

Dr Shaw relates that, among other curiosities taken from tombs discovered in the Sahara and which had belonged to women, he saw a piece of ordinary reed containing one of these pins and at least 30 grams of this dust.

In England, many people with gray hair have adopted a similar practice; but instead of using black in powder form, they use a kind of pencil in which the coloring matter is mixed with a fatty substance, like brown and black ointment sticks.

The question of whether hair is subject to sudden color change has often been discussed. Dr. Davy answered in the negative in a paper read at the *Bristol Association in* Manchester in 1861.

The general opinion is firmly in the affirmative, and several naturalists and physiologists conclude along the same lines. They cite examples of people whose hair has turned white or gray under the influence of violent emotions, such as pain, terror, etc. Haller reports eight such cases. Haller reports eight cases of such changes, but all he seems to admit is that this change can occur slowly under the influence of altered health. Marie-Antoinette has been cited by proponents of the general opinion as a striking and authentic example; but, when examined carefully, this case falls into the category admitted by Haller.

During her imprisonment by the Jacobins, the queen had been deprived of the use of the cosmetics with which she used to dye her naturally ashen hair black, and historians, in recounting her execution, repeat that her hair turned from jet-black to gray as a result of the moral torture the unfortunate queen had endured.

Had it been possible for a moral emotion, terror or grief, to suddenly turn his hair gray, surely the change would have been noticed before the time when the royal family was arrested trying to leave France. If such a metamorphosis were permissible, shouldn't we see it in soldiers engaged in terrible expeditions, amid the dangers and horrors of war? Dr. Davy himself had examined thousands of soldiers, prematurely exhausted by various climates, who had witnessed bloody battles and many of whom had received terrible wounds, and he declares that he has never come across a case of this kind.

Human hair cannot be injected. I have used coloring liquids such as silver nitrate solution and iodine solution, and have observed no change in color except in the positively immersed parts. Whether their color is due to a fixed oil, to a particular arrangement of the molecules of which they are composed, or to both, they are remarkably resistant to decomposition; they resist the action of acids and alkalis; only the strongest can dissolve them. They resist maceration and the action of boiling water itself, unless this long-continued action is combined with that of pressure, in which case they disintegrate and decompose. The sun whitens them, but this in no way explains the sudden change in color. Supporters of the popular view point to the changes that occur in the plumage of certain birds, such as the ptarmigan (Tetras lagopide, Tetrao lagopus L., Tetrao Alpinus Nils., Tetrao rupestris Gmel., or four-season partridge), and in the hair of certain quadrupeds, such as the mountain hare and the ermine, which turn white as winter approaches and take on a darker hue when it has passed.

A naturalist, Mr. Blyth, having examined a lemming killed in autumn, at the time when they change color, was convinced that the brown hairs had not changed, but had been replaced by new white hairs. In fact, there are reasons why the summer coat and plumage are darker in color than those of winter. The author concludes that it is impossible not to regard as erroneous the opinion that hair can suddenly change color under the influence of moral impressions.

Attempts by physiologists to explain such a change were unsuccessful, and more amusing attempts had been made to explain the phenomenon as something other than a deception. Doctor Davy, having taken up service on the continent, had occasion to know a military surgeon who had gone mad and to whom he was called fifteen days or three weeks after the invasion of the disease. The patient's hair, previously brown, had turned grey; but when the doctor called the attention of the regimental surgeon to this metamorphosis, the latter simply replied: "Your surprise will cease when you know that M***, since he fell ill, has stopped dying his hair."

The assassin Orsini, executed in Paris for having, in January 1858, made an attempt on the life of Emperor Napoleon III and mercilessly immolated twelve innocent people, exhibited the same anomaly. At the time of his arrest, the abundant curls of his hair were as black as ebony; but when he was led to the torture, they were iron-gray, simply because he had neglected the care of his grooming or because the cosmetics he used to dye them black had been removed... His friends and most newspapers naturally attributed this change to another cause, and we have no doubt that history presents this fact as the result of the overexcitement and moral suffering endured during his captivity.

As a general rule, you shouldn't dye your hair; it almost always harms one of the elements that together form the harmonious whole we call "physical beauty". The main elements of beauty, regardless of form, are complexion, eyes and hair. The first question to ask before attempting to change the color of such an important auxiliary must therefore naturally be this: will the change match the complexion and eyes? The Teutonic beauty of the Anglo-Saxons and Anglo-Normans has been passed on to the children of Britain, along with the practical good sense of some and the noble air of others. Most of the women whose charms form the ornament of England are therefore essentially blond, white and fresh, the very opposite of brown and black. Pale pink complexions, blue eyes and more or less chestnut hair dominate on this island. Now, to change the color of the face or that of the hair is to destroy the harmony of such a kind of beauty, because the eye cannot be changed accordingly; it is to produce an effect as unpleasant as that often produced by a badly dressed woman by a display of disparate colors in her toilette. Blonde people rarely benefit from dyeing their hair, if they ever do. Those who don't bear this characteristic sign of Teutonic origin, in whose veins the blood of a more southern race mingles, and whose dark complexion, gazelle eyes and jet hair all contribute to forming those beauties we call "brunettes", When youth begins to fade, or the curls of their hair turn silver and prematurely gray, they can call on art to help them restore their hair to its original color, without violating the principles of harmony. If the shade of hair is too bright, too vivid to match the eyes or the freshness of the cheeks, it can be softened by using a cosmetic that is sold under the name of "eau de brou de noix", but which is, in reality, nothing more than a solution of *potassium leadate*, prepared by dissolving a newly precipitated lead oxide in liquid potash until it is perfectly saturated.

2. Kohl

The word "kohol" comes from Hebrew and means "to paint". Women in the East used to and still do paint their eyebrows with various pigments; the most commonly used is antimony sulfate reduced to an impalpable dust. This practice has been introduced to some extent in England, but the kohol used does not contain antimony: it consists of a solution of Indian ink in rosewater. To prepare it, take a stick of Chinese ink weighing around 30 grams, pulverize it in a mortar - which is far from easy - then pour a half-liter of hot rosewater into the powder, and stir until the whole is equally liquid, a result only obtained after two days of trituration. The kohol thus prepared is applied to the eyelashes and eyebrows with a fine camel-hair brush.

3. Turkish Hair Dye

There are a few people in Constantinople, particularly Armenians, who are engaged in the preparation of cosmetics, and who make a great deal of money from those who wish to learn their art. Among these cosmetics is a black hair dye, which, according to M. Landerer (of Athens), is prepared in the following manner:

Galle nuts are pulverized, and a paste is then made with a little oil; this paste is cooked in an iron basin until the oil vapors stop escaping; the residue is then crushed and a new paste is made with water, which is put back on the fire to dry. To complete the preparation, a mineral mixture called *rastikopetra* or *rastik-yuzi* in Turkish, brought from Egypt to the markets of the Orient, is added. This foam-like metal, composed of iron and copper, is melted down by the Armenians on purpose. Its name derives from its use to dye hair, and particularly eyebrows, as *rastik* means "eyebrows" and *yuzi*, "stone". Reduced to a fine dust, it is incorporated as completely as possible with the gallnut treated as above, to form a soft paste that is kept in a damp place, where it acquires the property of blackening. Sometimes, this paste is mixed with odorant powders used as perfume in the seraglio and called *karsi,* i.e. "pleasant odor", the main component of which is ambergris. To use this dye, crush a little in the hand or between the fingers and rub it into the hair or beard. After a few days, the hair turns a magnificent black, and it's a real pleasure to see the beautiful black beards seen in the East among the Turks who use this preparation. Another advantage of this cosmetic is that the hair and beard remain soft and supple, and retain their black color for a long time after dyeing. It's safe to say that the coloring properties of this preparation are mainly due to the pyrogallic acid found in water treatment.

4. Eye and Mouth Cosmetics

This group of cosmetics won't stop us long. We have already mentioned kohol, which women in the East use to paint their eyebrows. In the name of hygiene, we reject the black dyes and powders that coquettes use to color the free edges of the eyelids and the corners of the eyes; women who resort to these means refer to them by a name almost as repulsive as the thing itself: "make-up".

Infection of the breath can have several causes: sometimes it is due to an alteration of the airways, more often to lesions of the teeth, tonsils and various parts of the mouth. In the first case, the bad odour can at most be masked by repeated aromatic lotions, to which a few drops of hypochlorite of lime are often added; but when the fetidness of the mouth is due to some local lesion, this infirmity can be cured by appropriate treatments and frequently applied aromatic gargles. Smokers and all those who have an interest in concealing the bad odor of their breath successfully use the small, aromatic, dry, hard, hard-to-melt *trochisques* known as "Bologna cachou".

5. Skin Cosmetics

Sometimes, we also try to brighten up the color of the lips, to protect them from chapping and cracking; *pommade rosat, rose ointment* or *lip ointment*, achieves this goal perfectly; but we must avoid using these strongly colored preparations such as *Psyche cream* or others, which for the most part contain zinc sulfate, lead acetate, etc., etc.

Physiologically, the skin is responsible for eliminating certain principles and absorbing others, and it's through the *pores* that this delicate dual function takes place. Its surface must always be clean and smooth. As an example of the dangers posed by any obstacle to its permeability, I cite the following experiment by Magendie:

"We coat a rabbit's body with a viscous coating, such as a concentrated dissolution of gum, gelatin and turpentine. These substances, innocent by nature, clump together the hairs and, as they dry out, imprison the whole animal, minus its face: the movements of its chest and the play of its main organs are unhindered; the skin alone no longer communicates with the atmosphere. The animal dies in a few hours, as if suffocated.

So, as soon as the skin's perspiratory functions are disturbed or suspended, the economy suffers. So it's perfectly within our rights to involve cosmetics, if only to avoid the fate of Magendie's rabbit.

Skin cosmetics are the most numerous, the most frequently used and the most justified by hygiene. Although the preparations recommended to make wrinkles disappear, to erase freckles, to redden or to color the skin in different hues, are most often the product of charlatanism, it is no less true that it is useful to maintain the freshness of the complexion, the smoothness, suppleness and elasticity of the skin, to fortify the tissues, prevent and dispel pruritus, loosen and remove epidermal debris, dissipate the odour of certain local sweats and, in a word, maintain the entire body surface in a constant state of cleanliness that enables the skin to perform its functions.

One of the first conditions for the proper preparation of skin cosmetics is that they should be free from any substance, poisonous or otherwise, which could attack or irritate the skin through contact, or which, if absorbed, would be capable of producing toxic effects; for if the absorption of poisons by intact skin in a bath can be questioned, the same cannot be said of alcoholic, acetic, glycerine or fatty preparations, etc., which are most certainly absorbed, either because they are permanently applied, or because they have the property of dissolving the coating covering the epidermis, which are most certainly absorbed, either because they are permanently applied, or because the vehicle has the property of dissolving the coating that covers the epidermis.

As skin cosmetics are very numerous, we divide them into waters, alcoholates and tinctures, vinegars, baths, emulsions, milks, pastes and flours, soaps, oils, ointments, glycerolates, blushes and powders.

6. Aromatic Waters

Aromatic waters are prepared by infusion, decoction or distillation; they are preparations that must be made at the moment of use, and cannot be preserved: they are emollient, soothing or astringent, depending on the substances they contain; moreover, they are little used in perfumery, apart from mint water, which is used to rinse the mouth, mainly after meals; but more often than not, artificial water, composed of mint essence and water, is used for this purpose. Other aromatic waters include rose, lemon balm, jasmine, orange blossom, cherry laurel, bitter almond and others. Both can be used for toiletries without any inconvenience.

7. Alcoholates or Tinctures

Alcoholates or spirits are formed from alcohol containing the volatile principles of one or more substances: they are simple when they contain only one plant, and compound when they contain several.

These preparations have varying degrees of alcohol content; the higher the alcohol concentration, the richer the resin, balsam or essence content. In this case, water clouds them. In the past, wine alcohol was the only alcohol used, but potato alcohol, beet alcohol or well-rectified grain alcohol can be substituted without inconvenience. This is what is done today.

Alcoholates, spirits or spirituous waters - in short, all the liquids used in perfumery whose vehicle is alcohol - are rarely used pure; more often than not, they are mixed with water, and sometimes adulterated with salts that precipitate on contact with this liquid. This is how a bad eau de Cologne containing *lead acetate,* which could cause serious accidents, is often sold on the streets. Alcoholic cosmetics are excellent and harmless; they cleanse perfectly, firm the skin, remove sebaceous secretion and perspiration products, and are harmless in all cases.

8. Vinegars

Cosmetic vinegars, like alcoholic tinctures, are prepared by maceration; they should be prepared with distilled or undistilled wine vinegar; but acetic acid from wood is used exclusively for this purpose, without any disadvantage; they are very rarely used pure, except when used as antiseptics; Most of them can be prepared by the consumer, who simply needs to choose a good vinegar. White wine vinegar is preferable, especially when it has been distilled, but wood acetic acid is most often substituted, with water added. In recent years, the pungency of these preparations has been reduced by adding a tenth of alcohol or a twentieth of glycerine.

Aromatic vinegars and all water-extended acids maintain tissue firmness, tone it up, correct its passive vascularity and varicose disposition; they clean the skin perfectly and act as astringents on mucous membranes. One of these aromatic vinegars is sold commercially under the name "Vinaigre des Quatre-Voleurs".

The fashion for these vinegars arose from the belief that they preserved those who used them from contagious diseases, an opinion no doubt born of the story of the *Vinaigre des Quatre-Voleurs.*

It is said that, during the Marseilles plague (1685 and 1721), four individuals, thanks to the use of this preservative, were able to approach a large number of plague victims without danger, and that, under the pretext of treating them, they stripped the dead and sick. Arrested later - the chronicle continues - one of them escaped the galleys by revealing the composition of this prophylactic.

We have banned vinegars from mouth hygiene. In this case, the acidic water coagulates the soap on the spot, in the pores of the skin and at the base of the hair. The fatty acids in the soap, thus set free, are no longer removed by the water: they become rancid and can produce sharp inflammations.

9. Aromatic Baths

Aromatic baths have recently become very popular. Aromatic alcoholates are most often used to prepare them, and can be combined with soaps and perfumed creams without difficulty. However, it is important to avoid combining the same soaps with vinegars, which break them down and set the fatty acids free. At other times, they are prepared by infusing or decocting aromatic plants, in which case they are considered tonic and stimulating; soapy, alkaline baths are sedative or resolving.

We move on to compounds which are all classified under the general heading of "emulsines", because all cosmetics included under this heading have the property of forming emulsions in water, i.e. giving it a milky appearance; they are generally oily solutions.

From a chemical point of view, they are an extremely interesting class of compounds, and very worthy of study. Because they decompose so easily, as can be seen from the way they are composed, they should only be produced in small quantities, or at least in quantities that can be disposed of quickly.

In store, they should be kept as cool and dry as possible.

10. Emulsions and Milks

The term "emulsions" or "milks" refers to a class of cosmetics designed to lotion the skin. They are formed by fatty or resinous bodies held in suspension in water by a mucilaginous, gummy or albuminous liquid.

But the name "milk" has also been given to the opaline liquid produced when resinous tinctures are added to water. This is how benzoin tincture is used to produce *virginal milk*, an often-used cosmetic that is said to erase ephelides or freckles, which it merely masks with a thin varnish on the surface of the epidermis. The skin inflammations sometimes caused by this milk, when perspiration is stopped, are even attributed to the formation of this layer.

Emulsive seeds, i.e. those with the ability to emulsify with water, such as almonds, pistachios, hemp seeds, etc., are used to prepare milks that are highly prized in perfumery, Unfortunately, they do not keep well, unless toxic salts such as corrosive sublimate are added, in which case they are veritable medicines that can only be sold by pharmacists. Examples include *Gowland's liqueur, Duncan's mercurial emulsion and Saemerling's cosmetics.*

Few perfumery items are easier to dispense than the cosmetics known as "milks". It has long been known that almost any kind of walnut or almond, shelled and skinned, reduced to a paste and ground with about four times its weight in water, yields a liquid that has every analogy with cow's milk. The milky appearance of these emulsions is due to the extreme division in water of the oil contained in these fruits. All emulsions, especially those made from bitter almonds and pistachios, are of great chemical interest, both because of their rapid decomposition and because of the products that result from their fermentation.

It is of the utmost importance to take the greatest care in handling the various milks intended for sale, otherwise these emulsions do not keep, and the loss is greater than the profit.

The elements of vegetable casein (contained in the seeds) are transformed, says Liebig, at the very instant when sweet almonds are converted into almond milk; this explains the difficulty many people find in making almond milk that doesn't spontaneously turn a day or two after it's been prepared.

Pure water is the cosmetic par excellence; but, although quite sufficient for those in good health, it often ceases to be so for city dwellers, whose health is rarely perfect, tested as they are by the cares of business, the heat of apartments, the defective ventilation of public buildings and meeting places, and by a sulfurous atmosphere saturated with the fumes resulting from the combustion of gas and coal. It is therefore necessary for art to come to the rescue of nature, from which we are too inclined to ask more than it can give.

Outside, as well as inside the home, on walks, at dances, in the evening, at public gatherings, in the midst of the day's vigils and various occupations, the skin of the face is soiled by dust that water alone would not remove. To restore its freshness, to correct the negative influence of city life, to give the complexion the radiance of health, no cosmetic comes close to rose emulsion. It purifies, softens the skin and makes it shiny, yet it's as harmless as an April dew on spring greenery. Eau de Cologne is also to be recommended.

By extension, the name "milk" has been given to compositions that become milky with agitation. Such is *antephallic milk*, a cosmetic containing sublimate, albumin, hydrated lead oxide, camphor and water. The original formula for this preparation belongs to Professor A. Hardy, a former physician at Saint-Louis Hospital. It is successfully used to treat ephelides or freckles. Composed of medicinal substances, it should be in the domain of pharmacy.

11. *Pasta and Flour*

Pastes and flours are used almost exclusively for washing hands. However, some people, notably the Russians, use them on their faces; they are made from almonds, pistachios, starch and more or less flavored cereal flours; when they are fresh, they present no danger in their use; but as they age, they become rancid and irritating.

Almond cake is most often used as a handwash, but is sometimes combined with other aromatic substances.

12. Soaps

Toilet soaps are the most important of all cosmetics; they were known in Pliny's time; the most esteemed came from Gaul; there were two types, soft and liquid. They were made with grease and lye from beech ash.

Medicated soaps containing sulfur, alkaline sulfides, iodine, etc., whose use has long been proposed and which industry has recently sought to exploit, are genuine medicines that should only be sold by pharmacists.

Toilet soaps can be hard, soft or powdered; all must be free of excess alkali, especially those used for the face, such as soap powders and creams, where this alkalinity is revealed by the causticity of their dissolution; it can also be recognized by treating a little soap with calomel (*mercury protochloride*); the mixture must not blacken.

Soapy lotions make it easier to remove perspiration residues, but when alkaline, they alter the epidermis, chapping and damaging the skin. Avoid using calcareous or selenite waters for beards.

Some soaps leave a smooth, velvety sensation; others, on the contrary, leave a harsh, dry sensation. These differences are due to manufacturing tricks, which we'll explain below. In *hot* soaps, boiling removes all traces of the caustic element from the paste. In *cold* soaps, an almost low temperature is maintained, even if this means leaving an excess of causticity in the paste. The former are the only hygienic soaps; the latter increase the seller's profits.

Indigo, aniline violet, caramel, chromium sesquioxide, turmeric, cinnabar (mercury disulfide), etc., are used to color soaps blue, violet, brown, green, yellow or pink; sometimes vegetable substances are used, which is much better; none of these substances is dangerous in soap; however, it would be better to replace cinnabar with another red.

It is unfortunate that the public is misled by soaps sold under assumed names which do not contain certain substances whose names they bear; it is especially unfortunate that therapeutic properties are attributed to them. Finally, we should mention a culpable fraud practised on soap powders, which consists in mixing them with 20 to 40% inert powders, such as *alabaster or talc*. These soaps are sold at very low prices.

Mr. Piesse has made a series of medicinal soaps such as sulfur soap, iodine soap, bromine soap, creosote soap, mercurial soap, croton oil soap, and many others. These soaps are prepared by adding the medicinal substance to tallow white, then tabletting. In the case of antimony and mercury soaps, the suboxides of the metals used can also be mixed in molten tallow white. Iodine, bromine, creosote and other soaps containing highly volatile substances are best made cold by grating the blanc de suif in a mortar and incorporating the medicines by a long trituration.

The author believes that in certain skin diseases, these soaps will be of great use as auxiliaries to general treatment. It is obvious that, during lotions, the absorbing vessels are very active; this type of soap should therefore only be used on the special advice of a skilled practitioner. It will no doubt soon be recognized that they can also be useful indoors. The precedent of Castile soap containing iron oxide suggests that these soaps will find their place in pharmacopoeias. Mr. William Bastick's discovery of the solubility in oil, under certain conditions, of active alkaloids, quinine, morphine, etc., makes it likely that analogous compounds can be made with soap.

Some forty or fifty years ago, several types of soap were imported that are now completely unknown: soaps from Joppa, Smyrna, Jerusalem, Genoa, Alicante and so on. Almost all were based on oil.

The sale of medicinal soaps, with sulfur, iodine, bromine, creosote, etc., which may be sold by perfumers in England, where pharmacy is free, cannot be tolerated in France; it is obvious that such preparations fall within the domain of pharmacy, and that whenever a soap or any other cosmetic is presented as possessing therapeutic properties, it is a *medicine* and not a cosmetic; it must therefore be subject in France to the legal prescriptions governing the sale of medicines.

13. Oils, Ointments, Glycerols

Aromatic oils and ointments are intended exclusively for the hair; however, some are especially useful for anointing the skin, cleansing it and making it supple. *Cucumber ointment* and *cold cream* are excellent preparations when fresh, but should never be allowed to go rancid, and should never be used in excess.

Glycerols, or flavored glycerins, may soon take on an important role in perfumery, but for this to happen, they must be used in their purest form. Glycerine is soluble in water, does not go rancid, dissolves almost all substances dissolved by water and alcohol, is soft and creamy, and has, in a word, all the advantages of fatty substances, without the disadvantages.

Glycerine is widely used to treat chapped lips, despite the fact that it is sticky.

Glycerin soaps" are a speciality of soaps widely used for their skin-softening properties.

Glycerine lotions are often preferred to soaps, as it's not uncommon to find foreign adulterators sending us soaps with not even a trace of glycerine under the name of "glycerine soaps".

14. Blushes

Whatever form they take, *powder*, *paste or crepe*, whatever material they're made of, blushes are the most dangerous of cosmetics: sometimes they obstruct the skin, making it hard and brittle and preventing perspiration; sometimes, as a result of their absorption, they cause veritable poisoning.

Ovid indicates various artifices to correct nature: "You borrow from ceruse its deceptive whiteness; other artifices replace the color of blood; you know how to lengthen or thicken your eyebrows and erase under a cosmetic your true cheeks; you are not ashamed to enliven the brilliance of your eyes with fine powders or with saffron that grows on the limpid banks of the Cydnus" (Ovid, *Art d'aimer*, III.).

Cumin was said to make people pale. Pliny tells us that mandrake was used to erase facial scars; according to Ovid, poppies were used for the same purpose.

White blushes, prepared with ceruse (lead carbonate) and disguised as "silver white", "pearl white", etc., are the most dangerous. Poisonings caused by these make-up products are numerous, especially among dramatic artists. In addition to the disadvantage of absorption, they alter the skin, cauterizing it and irritating it chronically; they impart a sallow tint and a wrinkled appearance, due to a loss of retractility and reduced capillary circulation.

Absorbed through the skin, in whatever form it is applied, lead passes into the bloodstream, where, instead of spontaneously revealing its presence through some awakening seizure, it operates mutedly and slowly. Its first effects are felt in the nervous system: strength is depleted, sensibility is perverted or exalted; then there are contractures, spasms, automatic movements, epileptiform convulsions, and often even signs of softening of the medulla or brain.

Colic, encephalopathy and lead paralysis are the consequences of frequent use of these blushes. Only recently, Dr. Ward published an observation of lead paralysis in several members of a family using lead-containing crepons.

To replace ceruse, we've tried bismuth sub-nitrate, which is the true white of blush, zinc oxide and oxalate, chalk, talc, etc.; but, as ceruse adheres and covers better, we prefer it, without thinking of the dangers to which its use exposes us.

Too much attention has been paid to the traces of arsenic that bismuth sub-nitrate may contain; first of all, it would be easy to obtain it pure; then, arsenic being present in an insoluble state and in infinitesimal quantities, it would be completely harmless.

Red blushes are prepared with the same materials and colored with carmine, Brazil wood and carthamine: all these substances are harmless, but cinnabar is at least as dangerous as ceruse.

Blue blushes owe their color to indigo and azure blue, while gray blushes owe their color to antimony sulfide; they have no disadvantages other than those resulting from the presence of ceruse, and those presented by all these coatings.

15. Powders

In therapeutics, old wood powder, lycopod, starch or starch, pure or sometimes mixed with bismuth sub-nitrate or calomel, are used as absorbing and drying powders to prevent the parts from coming into contact with each other; in perfumery, *flavored starch* and *rice powder are used* above all; their use presents no disadvantages.

Rice powder refreshes and softens the skin while absorbing moisture. However, this powder usually contains starch, talc, alabaster (sometimes 50%) and carbonate of lime, flavored with a hint of violet. It's even better for what's known as "fleur de riz", in which case rice often doesn't enter at all; it's sometimes replaced by magnesia, to give the mixture greater lightness and suppleness.

If no more serious substitutions were ever made, we shouldn't complain too much, because the resulting compound is harmless and, what's more, achieves its purpose much better than the powder itself. The aim is to make the skin appear whiter and finer. Genuine rice powder, however, offers too little fixity, and would be removed by the mere brushing of air or fabrics; on the contrary, the powder that is sold under this name can, for an entire evening, put up the most magnificent resistance. Unfortunately, to make it even more stable, some manufacturers add astringent powders whose action can cause serious accidents.

16. Sponges

The best sponges come from Smyrna or the shores of the Greek islands. When we receive them, they are full of sand; it's in this state that it's best to buy them. We then beat the sand out of them with a stick, and rinse them in cold river water. Nothing is better than a good sponge for cleaning the skin, and surgeons prefer it to any other substance. When sponges are used with soap for regular washing, they soon become greasy and are discarded before they are half used. The cellular fibers that make up the sponge's particular fabric make it suitable for breaking down soap, retaining grease and oil and making it *sticky*. When this happens, prepare a soda solution in the proportion of 250 grams of soda to 2.5 liters of water, and let the sponge soak for twenty-four hours. Then wash and rinse in spring water, followed by water containing a little muriatic acid (one glass of acid for 2 to 3 liters of water is sufficient). Finally, rinse the sponge once more with plenty of spring water. If you take care to rinse your sponge thoroughly each time you use it, you'll rarely need to do this.

17. Tattoos

Before Caesar conquered the Gauls, the Picts, the first inhabitants of the Scottish mountains, took their name from the colors they were covered in; Papdïcas chiefs and nobles wore their coats of arms tattooed on their foreheads and chests.

In Europe today, tattooing is reduced to covering the arms and chests of our sailors and workers with emblems and mottos.

The drawing to be represented is first traced with a feather or brush on the area to be tattooed. A red, blue or black colorant is diluted in a vase, on a palette or in a shell, as if painting. Two, three or four sewing needles are attached together and abreast. The skin on which the drawing has been traced is stretched as evenly as possible, and the tattooist, after dipping his needles in the colored solution, pushes them into the thickness of the dermis, following the contours of the image. The needles are not placed in the same direction as the lines, but across them; for it is not to spare time and pain that they are brought together several times; it is to give more width to the lines, and thus make each point prick several times to make it more apparent.

The needles are inserted more deeply into the dermis, depending on the thinness of the skin, the sensitivity of the patient, or the will of the engraver who, with each new puncture, dips his burin back into the colored liquid, if he wishes to act conscientiously. The operation is complete when the entire drawing has been punctured. After a quarter of an hour, the tattooed person washes off the area that has oozed a little blood, sometimes with water, sometimes with urine: some artists prefer brandy or rum, of which there is enough left in the glass for them to make a profit.

Sometimes, out of an abundance of caution, a swab or finger impregnated with the coloring material is passed over the pricks of single-color tattoos, to allow more of it to penetrate the skin. But this is hardly necessary, and would be impossible for differently colored designs.

The most commonly used dyes are: *vermilion, crushed powder, Indian ink* and *blue,* used by laundresses, diluted in pure water or saliva, and black or blue *writing ink.*

The inhabitants of Oceania and the natives of America tattoo their whole body, especially their face; the Germans dyed their skin red. The use of these paints and greases was a hygienic necessity for the peoples who used them; they protected themselves from the sting of insects, and made themselves less sensitive to harmful fumes and atmospheric changes.

In Florida, women cover their entire bodies with indestructible designs; those from Duan have flowers of different colors engraved on their skin. The inhabitants of Rotouma and the Wallis Islands have their lower chest, up to the knee, covered with a regular tattoo imitating the thighs of the ancient knights.

In Tahiti and the islands of Oceania, the ears of the inhabitants are pierced and wear flowers and fragrant herbs instead of pendants; Cochinchin women blacken their teeth by using betel; Moresque and Tunisian women dye their cheeks and lips with gallnut, saffron, henna, etc.

After this enumeration of such varied cosmetics and such diverse practices, we will formulate a precept:

Whenever we use artificial means to preserve the radiance and glow of the complexion, all the attributes of external beauty, it will always be at the expense of general health. A well-ordered diet, sobriety and moderation in all things are the safest cosmetics.

18. Perfumes, Medicines and Poisons

Fragrances, perfumes and essences are sometimes used medicinally. In addition to musk, castoreum, civet and amber, which are frequently used as antispasmodics, we could mention the gum-resins of the Umbelliferae (*assa foetida, galbanum, sagapenum, opoponax* and *gum ammoniac*), to which the same properties are attributed, and the fragrant resins of Conifers and Terebinthaceae, such as frankincense, bdellium, elemi resin, myrrh, animated resin, galipot, etc., which are often used as antispasmodics., which are used externally as maturatives and resolvatives, and form part of a large number of plasters and ointments. Balms from Tolu, Peru, Mecca and benzoin are excellent expectorants as well as exquisite perfumes.

Especially in recent years, therapeutics have taken great advantage of the fact that most perfumes volatilize when heated, to create fumigations and even inhalations: fumigations of aromatic plants, juniper berries, etc., exhalations of essential oils of tar, and benzoin vapors have rendered great service in aphonia and voice extinctions.

Essential oils themselves are often used in medicine, sometimes pure internally, such as aniseed, bitter almond, turpentine, sabine, copahu, etc., and sometimes externally, in the form of alcoholates, making them excellent fortifying, rubefacient and derivative medicines. Lastly, essential oils form the basis of the distilled waters or hydrolats so frequently used, and among the most fragrant are those of bitter almond, cherry laurel, aniseed, fennel, orange blossom, cinnamon, lemon balm, sweet clover, lime blossom, roses, etc.

As for the emanations of plants, flowers and fruit, we need to distinguish two cases: sometimes the scents they give off are due to essential oils, to more or less coercible perfumes; other times this simple release of odors is complicated by chemical phenomena that purify or alter the ambient air, depending on the circumstances in which they occur.

The odoriferous emanations of flowers, or those of isolated perfumes, are not always harmless; they must always be breathed in moderation and diluted in large quantities of air; they often produce intense cephalalgia; and some distillers, and especially workers who manufacture artificial essences (compound ethers), often suffer from quite serious nervous disorders; in addition to this, there are special idiosyncrasies that make certain otherwise very mild odors repel.

Under the influence of solar radiation, all green parts of plants purify the ambient air, absorbing carbonic acid, fixing carbon and releasing oxygen; in the dark, on the other hand, these same green parts exhale carbonic acid, and this same exhalation is constantly noticeable in all the colored (non-green) parts of plants, such as corollas, stamens, pistils, ripe fruits, etc. M. Boussingault's research has even shown that, under certain circumstances, plants can release carbon monoxide, which is a violent poison. We must therefore conclude that it is never wise to leave flowers in large quantities in enclosed apartments.

Some essences, even when exposed to air, oxidize, resinify and release carbonic acid, and some can ozonize the air, giving it new properties.

In addition to the beneficial and salutary uses of perfumes, we should also mention the harmful and criminal uses of perfumes; for, if they are used as medicines, they are also used as poisons, or rather they are both.

At various times, sorcerers, soothsayers and enchanters have hidden their mysterious practices under the veil of perfumes; however, it would be very difficult for us to say what is this subtle poison that kills by simple inspiration, which the Italians kept in the bezels of rings, and which René the Florentine is said to have used to poison Jeanne d'Albret.

Famous poisoners made great use of perfumes. Medea, well versed in their knowledge, had recognized the beneficial effects of bathing in aromatic vapors: this was what constituted her magical power. M. Beleo tells us that she was the first to discover a flower that could make hair black or white, so that those who wished to change the color of their hair had their wishes fulfilled thanks to her; to prevent doctors from penetrating her secrets, she prepared her baths mysteriously. Fomentations restored strength and health to men; and, as she used a boiler of wood and iron, it was believed that she actually boiled her patients. Pelias, a cacochymous old man, submitted to her prescription and died in the course of the treatment.

This chapter concludes our remarks on the application of fragrances to the hygiene and grooming of elegant beauty.

To be "in good odor" is an indication of moral purity. Dr. Andrew Vinter would have each person adopt a special scent, in the physical sense of the word, according to the circumstances of age, joy and sorrow in which they find themselves. Why, he says, shouldn't we recognize our beautiful friends by the delicate fragrances that surround them, just as we recognize them from afar by the sweet sound of their voices? For each character, there is a scent that seems to belong particularly to her. To the spiritual woman, jasmine; to the brilliant woman, magnolia; to the strong woman, musk; to the young girl in the first flower of her beauty, rose. Lemon's emanations are better suited to melancholic natures, and heliotrope has a sad note that suits the young widow.

The sovereign Creator not only wanted all his works to be useful, he also wanted to give them beauty and variety. Flowers could have all had the same color and smell, they could have been odorless and colorless. And yet, what exquisite beauty, what diverse fragrances in plants! And how we all admire these brilliant hues, these embalmed emanations! Man is made to appreciate the gifts that creative goodness has sown in his path with such profusion, and the pleasure he derives from them, as it is the purest and most innocent, is at the same time the sweetest and most lasting.

"Solomon in all his glory was not dressed like one of these flowers," said the divine Master, speaking of lilies; and, when he wants to give an idea of his greatness and glory, it is to a flower that he compares himself: "I am," he says, "the rose of Sharon."

IV. GENERAL FRAGRANCE APPLICATIONS

1. Perfume Boxes

The earth smiles in all its glory,
Here, fragrant plants exhale their perfumes.
Creation by Joseph Haydn.

The perfumes used by the ancients were certainly none other than the fragrant resins that flow naturally from various trees and shrubs in the Eastern hemisphere; to convince us of their use and purpose, we need only read the Scriptures: "Who is he that cometh... perfumed with myrrh and frankincense with all the powders of the merchant..." (Song of Solomon, III, 6). "(*Song of* Solomon, III, 6.) To abstain from the use of perfumes is regarded in the East as a sign of humiliation. "(Isaiah, III, 24.) "And they came and brought tablets" (Exodus, XXXV, 22.) The word "tablets" in this passage means metal, wooden or ivory perfume boxes, curiously inlaid. Some of these boxes were perhaps made in the shape of buildings, which would explain the word "palace" in Psalm XLV, 8: "All the garments smell of myrrh, aloes and cassia, whose odor escapes from the ivory palaces by which they have made you joyful." From what is said in St. Matthew (II, 11), it would seem to follow that perfumes were considered the most precious gift possible: "And when they (the Magi) had opened their treasures, they presented him (the infant Jesus) with gold, frankincense and myrrh." As far as we know, the perfumes used by the Egyptians and Persians in the early days of the world were dry perfumes, nard (*Nardostachys Iatamansi*, Valerianaccae), myrrh, frankincense and other resins, almost all of which are still used by perfumers. Among the curious objects on display at Alnwick Castle is a vase found in the catacombs of Egypt. It is filled with a mixture of resins, etc., which still give off a pleasant scent today, although they are probably three thousand years old. It is certain that this vase and the preparation it contained were used to perfume apartments, just as potpourri is used today.

2. Powders for Sachets

Perfumers in France and England prepare a large number of these powders, which are easily sold in silk sachets or elegant envelopes. These sachets, whose scent we like to breathe in, also provide an economical means of imparting a pleasant fragrance to linen and clothing, when left in drawers.

3. Spanish Skin

Spanish skin is a highly scented leather that is soaked in a mixture of essences in which a few fragrant resins have been dissolved.

These skins exude a very pleasant scent for years, which has often led to them being called "inexhaustible sachets". Thin ones have been made and are much sought-after for scenting stationery.

The permanent scent of Russian leather is well known and appreciated by many. It is due to the fragrant sandalwood with which it is tanned and the empyreumatic oil of birch bark with which it is wrought. But the smell of Russian leather is not sufficiently *sought-after* to be considered a perfume; however, by dipping it in the various essences, it can be given all possible odors, and it will retain them in a remarkable way, especially if dipped in the essence of sandalwood or schoenanthe. In this way, the scent of Spanish skin can be varied ad infinitum, and its merit greatly increased by the fixity given to the perfume.

4. Scented Stationery

If a piece of Spanish skin is placed in contact with paper, it will absorb enough of the scent to be considered "perfumed". It goes without saying that, in order to be able to write on paper, none of the fragrant dyes or essences must touch it, as these substances would alter the fluidity of the ink and hinder the movement of the pen; it is therefore only by means of this kind of contagion that it is possible to perfume writing paper with advantage.

After the sachets we've just mentioned, we should mention the scented wadding used to line all sorts of objects used in ladies' boudoirs. It is used in pincushions, jewelry cases and the like. To prepare this cotton, simply soak it in some strong tincture of musk, etc.

5. Scented Bookmarks

We saw in the manufacture of Spanish hides how leather can absorb odorous substances; this is absolutely the same way cards are treated before being made into bookmarks. Thus prepared, they are then decorated with designs to suit the buyer's taste, sometimes with embroidery, sometimes with pearls.

6. Scented Stones

We're curious to know how these stones can smell like natural flowers, and where they come from.

When you move them around in the little box containing them, you see the curious effects of the kaleidoscope and breathe in the most delicious perfume. The truth is that, under the silver paper on which the stones are fixed, is a card cut to the size of the box; on each card is spread a mixture of musk, civet and rose essence crushed and mixed in a little tragacanth gum.

7. Cassolettes and Spring Dishes

These are small ivory boxes of various shapes, pierced to let out the scent they contain. The mixture used to fill "these ivory palaces that make us joyful" is made up of equal parts of grainy musk, ambergris, vanilla seeds, essence of rose, iris powder, with enough tragacanth to give the whole the consistency of a paste.

8. Scented Shells

The Venetian shells found in such abundance on the shores of the Adriatic Sea, near the islands of Greece and the Maldives, are first cleaned with weakened muriatic acid to give them their pearly sheen. A mixture of essences is then made, for example 500 grams of bergamot and 25 grams of sandalwood, 56 grams of lavender and 56 grams of rosewood, with 2 grams of civet and 3 or 4 grams of musk.

The shells are then dipped in the composition that rises in the spirals of which they are composed. When dry, the shells are used to perfume jewel cases and work boxes.

9. Fragrant Flowers

Artificial flowers were so successful, however, that they only gave the illusion to the eye, so the perfume industry hastened to add the scent of the imitated plant to its appearance.

So we now have plants, flowers and bouquets whose freshness and fragrant qualities are absolutely similar to the natural ones.

10. Scent Jet Rings

In addition to perfumes for hand-kerchiefs, clothes and other items of clothing, there are a host of small scented objects that are becoming increasingly popular.

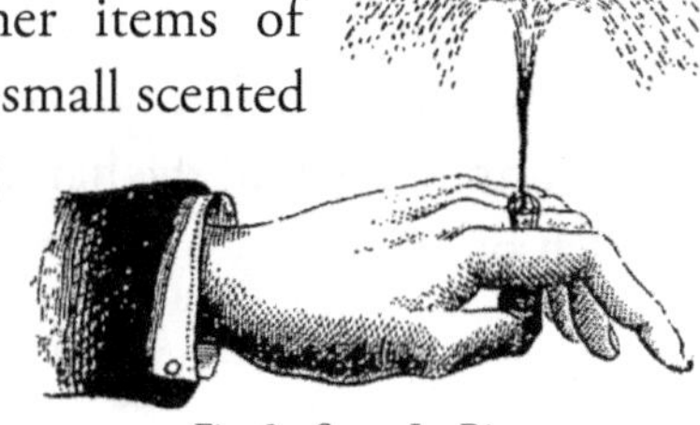

Fig. 3 - Scent Jet Rings.

At the top of the list is the scent-jet ring. This ring once enjoyed a kind of celebrity status as a means of carrying scents on one's person.

The success of this charming jewel with all those who have seen it could not be more flattering for its inventor. It is an object of both luxury and utility. With the lightest pressure, the wearer can release a jet of perfume at will. So anyone can carry enough essence to freshen up a moment at a ball, a concert or in a sick room.

More than once, these rings have been the occasion for comic scenes. For example, a gentleman who abhors perfumes, except tobacco, pressing a lady's hand receives a shower of the eternal frangipane or the no less eternal *kiss me quick*, much to the delight of the company, delighted to see him so gently "unveiled".

These rings are very easy to fill. The bezel of the ring is squeezed; scent is poured into a cup and the ring dipped in it; the elasticity of the bezel draws the scent inside until it is filled.

11. Scent Jet Pins

In the same genre, and with equal success, we created the magic tie.

The pin of this tie is hollow and communicates by means of a hose with a rubber bulb concealed under the clothing.

A simple touch of the hand sends the scented essence to a given point.

12. Incense

There's no doubt that the habit of burning pastilles in apartments stems from the custom of burning incense on altars during religious ceremonies. - (Luke, I, 9.) "And it came to pass by lot, according to the custom among the priests, that it was for him (Zacharias) to enter the temple of the Lord to burn incense there.) "And you shall make an altar to burn incense on... And Aaron shall burn incense on it every morning when he arranges the lamps; and in the evening, when he lights the lamps, he shall burn incense on it." (Exodus, XXX, 1 and 7.)

Fig. 4 - High priest burning incense on the altar.

13. The Censer

The censer used in churches is made of copper, German silver or other precious metals; the engraving shows its shape; the upper part is pierced with holes to let the perfume evaporate. (The word "perfume" is derived from the Latin *per fumus*, "by smoke", because the first perfumes in use were those whose odor is released by combustion). In the lower part is a small copper stove that can be pulled out and filled with burning coal. To use it, place the charcoal in the censer and pour the incense over it: the heat immediately

Fig. 5. - Censer.

volatilizes it and the smoke spreads. The cleric, by swinging the censer attached to three long chains in the air, encourages the release of the embalmed vapour. The manner of incensing varies slightly in the churches of Rome, France and England; some send the censer higher than the head. At the Madeleine church in Paris, it is customary to throw the censer from the length of the chains and to catch it quickly with the left hand.

The engraving depicts an antique incense box or burner (fig. 6.); the silver original is 18 centimeters long. It belongs to William Wells, esq. of Holme Wood House, Whittlesea, Cambridgeshire. It was found while cleaning the Whittlesea pond. Its shape and construction are well matched to its intended use. When not in use, it is an elegant object to adorn a boudoir; and, when needed, incense and matches to burn it are found in the box. It is probable that it belonged to Ramsey Abbey (*Ram*, in English "ram"), a supposition suggested by the ram heads seen at the front and rear of the vessel.

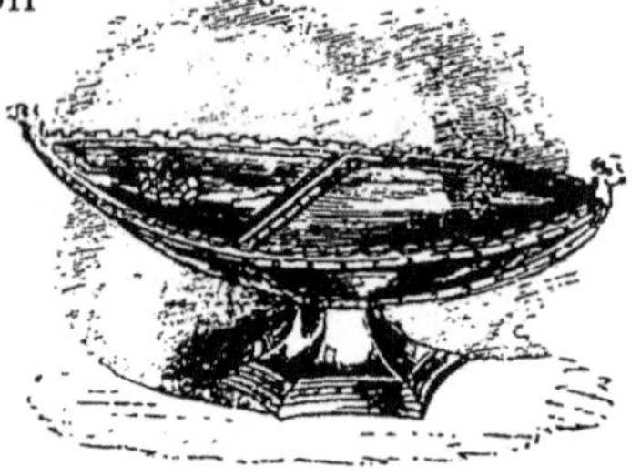

Fig. 6. - Brûloir à encens antique.

Various samples of incense prepared for the altar service appear to be no more than ordinary olibanum resin, which bears no resemblance to the composition prescribed by God and whose formula is given at length in Exodus.

Modern pastilles are really just a slight variation on the incense of the ancients. For a long time, they were known as "Cyprus knucklebones". In old pharmacy books, under the name of *suffitus*, we find a certain mixture of resins known at the time, which, when thrown on hot ashes, produced a smoke considered salutary for many illnesses.

It's with the same idea in mind, or at least to mask the bad smell in patients' rooms, that fumigation tablets and ribbons are used today.

14. Perfume Lamp

Shortly after the discovery of the peculiar property of platinum sponge to remain incandescent in alcohol vapour, J. Deck (of Cambridge) made an ingenious application for perfuming apartments. An ordinary spirit-wine lamp is filled with Hungarian water or other scented spirit, then a wick is placed in it as usual. In the center of the wick, about 3 millimeters above the wick, is a small spongy platinum ball attached to a small glass rod inserted into the wick.

Fig. 7. - Flameless fragrance lamp.

When the lamp is set up in this way, it is lit and left to burn until the platinum turns cherry red; the flame can then be extinguished, and the platinum continues to glow for an indefinite time. The proximity of a burning ball naturally leads to the evaporation of a volatile body, such as an aromatic essence spread over the surface of a cotton wick, and consequently to the diffusion of the odor.

Instead of filling the lamp with Hungarian water, you can add Portugal water, verbena water or any other spirit essence.

People who are accustomed to using perfume lamps often notice that, whatever the difference in the composition of the liquid used, there is always a certain resemblance in the odor when the platinum is in use. This is because, as long as the alcohol vapour mixed with oxygen gas passes over the incandescent platinum, various products are always formed, in greater or lesser quantities, such as acetic acid, aldehyde and acetal, which give the vapour a particular cachet and a rather pleasant bouquet, but which absorb and annihilate all other odours.

15. *Fumigation Paper*

There are two ways to prepare it:

A. Take sheets of light cartridge paper, dip them in an alum solution made as follows: alum 28 grams, water 1.56 liters. When they are completely soaked, dry them well; on one side of the paper, spread a mixture of benzoin gum, frankincense and Tolu or Peruvian balsam in equal proportions, if you prefer

Fig. 8. - Vase and section showing the Bruges ribbon.

benzoin alone. To spread the resin, etc., it is necessary to melt them in an earthen vase; then pour a thin layer onto the paper and smooth the surface with a hot spatula. When you want to use it, hold strips of this paper over the candle or lamp to evaporate the odorant without setting the paper ablaze. Alum prevents the paper from catching fire to a certain extent.

B. Leaves of good light paper are soaked in a solution of saltpetre in the proportion of 56 grams of saltpetre to 56 centilitres of water; they are then dried thoroughly.

A fragrant gum, myrrh, frankincense, benzoin or other, is dissolved in rectified wine spirit until the latter is saturated with it; with a brush, this solution is spread over both sides of the paper, or the paper is dipped in the solution; after which, it is suspended in the air, where it dries rapidly. Strips of the paper are then rolled into spits, which are ignited and immediately blown out. The nitre contained in the paper causes it to burn gently, spreading the pleasant fragrance of aromatic gums. If you press two of these leaves together before the surface has dried, they will stick together and form a single sheet. Cut into strips, they form what are known as "scented matches" or "scented brooches".

16. Vaporizers

For some time now, certain devices called "vaporizers" have been widely used, which, by means of a high-speed air current, disperse perfume-laden alcoholic liquids into the atmosphere in the form of vapors.

Two of the most common types of vaporizers are mentioned here:

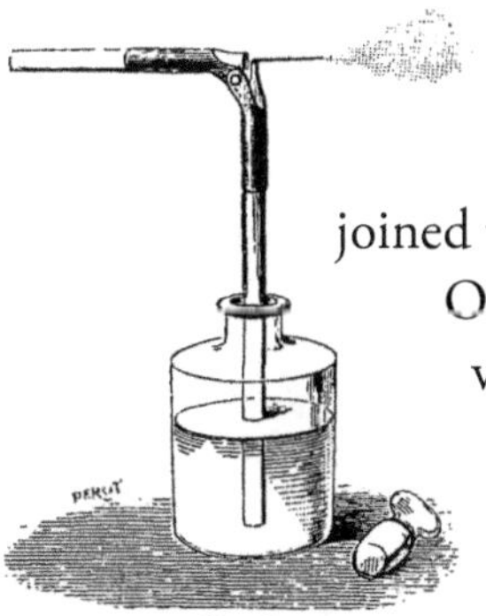

The first can be used on all bottles containing the liquid to be sprayed.

It consists of two glass tubes joined together by a movable hinge (fig. 9). One of these tubes ends in a cone with a very small opening.

To use this device, place the tubes at right angles to each other, dip the conical-ended tube into the bottle so that the lower end is at a small

Fig: 9.

distance from the bottom of the bottle; blow strongly into the horizontal tube, and the current of air, carrying the liquid contained in the bottle through the vertical tube, is mixed with the vaporized alcohol which serves as a vehicle for the perfumes contained in the bottle.

The second device consists of the liquid bottle and the spray system. Here, the intended purpose is no longer achieved by entrainment, but, on the contrary, by pressure on the liquid, which rises in the dip tube and is mixed with a certain quantity of air when the rubber ball at the top of the device is compressed.

This system is preferable to the previous one: it is more convenient and gives a more divided spray.

Fig: 10.

THE POLITICAL ROLE OF FLOWERS

We're all familiar with the language of flowers, which, in the days of Florian's shepherds and shepherdesses, expressed themselves with such naivety and admirable discretion. But when flowers get involved in politics, their equally mute way of speaking is just as expressive. Unfortunately, it no longer has amiability as its object; instead of uniting, its mission seems to be to divide. So flowers can become the victims of the political passions that reign in our modern societies.

In all countries of the civilized world, changes of government bring about a veritable hecatomb among administrative officials; some arrive on the scene, others leave through the wings. The same is sometimes true of ornamental plants, when they become emblems of the policy of the day; others, on the contrary, are signs of opposition. Experienced administrators, like beautiful plants, have often been sacrificed to the passions of the parties that have divided our country.

I'm not going to talk about the War of the Two Roses, or the graceful emblems under which two political parties fought bloody fratricidal battles. I'll just mention, in a few words, the flowers that have played an important role in the history of our beautiful France.

General Fragrance Applications

The white lily - if it is the white lily, and not the bee[1], that was originally depicted on the flag of legitimacy - originated in the East and is found in a spontaneous state in Palestine[2]. It has probably only been known in France since the time of the Crusades, along with a number of ornamental plants that can still be found today on the ruins of old feudal castles, where they have become naturalized. The white lily commends itself to the eye through its elegant habit, its beautiful snow-white corollas, and to the organs of olfaction through its sweet scent. Today, it is rarely cultivated, except by those of our farmers who attribute somewhat problematic therapeutic virtues to its onions. It is generally preferred to several other species of the same genus which, while not inferior in beauty, have more or less odorless flowers. Horticultural fashion loves novelties, and like politics, it has its whims and demands.

Under the Restoration, the heraldic lily had an adversary. Many orangeries grew the *Geranium tricolor* Andr. which, from an aesthetic point of view, was far from its merit; but it was far superior to it as an emblem of opposition. This is so true that, once the flag of the Revolution

---

1. In 1653, the tomb of King Childeric I was discovered in Tournay. Among the various objects found in this tomb were several gold bees. The fleurs-de-lis of the kings of France resemble a bee, seen from the back and upside down, rather than the six-petal Lilium candidum flower. On this contentious issue, see Leber, Collection des meilleures dissertations, notices et traités particuliers relatifs à l'histoire de France, Paris, t. III (1838), pp. 168, 198 and 292.

2. Linné (Sp. Plant., 433) mentions it in Palestine and Syria. It was known in Solomon's time, where two lilies were engraved on the basin known as the Brazen Sea, located in the Jerusalem temple (Bibl. sacr. Regum, III, VIII, 26).

and the first Empire had been re-established on our public monuments, this *Geranium* disappeared to such an extent that I haven't seen it since. Its reign was over; it no longer had a raison d'être.

However, it did have its enemies. I've heard it said that, shortly after Louis XVIII's return to France, a lady of high nobility, having ridden to her newly acquired château, dismounted in front of the greenhouse, entered it and, on seeing a large collection of this revolutionary plant, tore it to pieces with her riding crop. This is how politics hates innocent plants and punishes them with proscription.

I can cite another similar example from the same period. A gardener, a good royalist, in charge of cultivating a garden and looking after a greenhouse, came across, *horresco referens*, a collection of *Hydrangeas*, introduced by his predecessor who had, no doubt, sacrificed himself to the star that had just faded and disappeared. Without asking the owner's permission, this plant disappeared without a trace, not even, it is claimed, in the pocket of this reform-minded gardener. Why this summary execution? The cause lay in the ignorance of both the person who had introduced it and the person who had exterminated it. Both assumed that the name had been given to the plant in honor of Queen Hortense. This attribution is absolutely impossible, as we shall demonstrate.

Was it the first time that the name of this queen of Holland appeared in history? A famous Roman orator, *Quintus Hortensius,* born in 113 B.C., had long been

General Fragrance Applications

known. Could it be him, by any chance? Why not? Botanists have named a North American tree *Virgilia, and* a South American tree *Theophrasta.* But there was a serious reason for honouring these: Virgil's name recalls the *Georgics,* and Theophrasta's a book on the causes of vegetation. The orator Hortensius recommends himself only by rhetorical flowers, and this is not a scientific title.

If you open Bouillet's *Dictionnaire d'histoire et de géographie,* in which so much is to be found and which is available to everyone, you will read, in the article "Commerson", that this botanist travelled around the world and collected the richest herbarium known up to that time. This herbarium has been deposited at the Muséum d'Histoire Naturelle in Paris. It contains the plant we're talking about, labelled *Hortensia* in his own handwriting. It was spread and cultivated under this name, which was imposed on it by the author of its introduction to the Île-de-France and the gardens of Europe. Commerson died in 1773, long before Queen Hortense, born in 1783, was mentioned.

The person to whom this plant was dedicated is well known. It's not a queen, nor a princess, nor an orator, nor a poet, nor even a learned naturalist. It's simply the name of a good Parisian bourgeois, Hortense Lepaute, wife of a renowned watchmaker, of whom Commerson was a friend.

We can therefore grow this beautiful plant in the Republic in all good conscience, without casting aspersions on any political party. It originated in China and Japan, and is frequently depicted in Chinese and Japanese paintings. It is passionately cultivated in both countries, where horticulture is highly prized. We believe that we have definitively rehabilitated the *hydrangea* and calmed minds about it.

Finally, the violet has more recently been chosen as the emblem of a political party, which in our opinion is a real usurpation. From time immemorial, this commendable plant has belonged to an infinitely respectable party, which only deals with the internal politics of the house, and generally with skill. It almost has the authority of numbers, since it constitutes the most amiable and influential half of the human race. The humble violet has the double merit of blooming its corollas in the first spring and spreading its sweet perfumes all around it, when very few other plants have reached the time of their flowering. Scientists have also distinguished it, giving its name to one of the colors of the solar spectrum. The bouquets we make of them are the delight of the ladies of our cities, and especially of Paris, where there is a considerable trade in them. In the countryside, young girls also appreciate these little spring flowers, happily gathering them to adorn themselves on Sundays and feast days.

We conclude from all these facts that flowers are the emblem of peace, that their mission is to charm man and especially his companion. It would therefore be a mistake to choose them as signs for political parties, since these

are constantly fighting each other by word, by pen, by
the press and, unfortunately, sometimes even by more
violent means, as history shows, even among peoples
who consider themselves to have reached a high degree
of civilization.

Dominique-Alexandre Godron
(Extract from Mémoires de l'Académie de Stanislas, 1878).

TABLE OF CONTENTS

THE LOST ART OF PERFUME 7

I. PERFUMERY THROUGH THE CENTURIES ... 9

1. Jewish Perfumes 10

2. Chinese Perfumes 12

3. Perfumes in the Orient 12

4. Scythian Perfumes 13

5. Perfumes in Egypt 13

6. Greek Perfumes 13

7. Roman Perfumes 18

8. Perfumes in Italy 20

9. Perfumes in England 20

10. Perfumes in France 23

II. SMELL AND ODORS 30

1. The Sense of Smell 32

2. Smell Theory 33

3. Odor Harmony 38

4. Land Odor 44

5. General Air Odor 45

6. Variability of Fragrance Qualities 47

General Fragrance Applications

III. FRAGRANCE AND COSMETICS HYGIENE 53

   1. Hair Hygiene, Depilatory Preparations 60

   2. Kohl 67

   3. Turkish Hair Dye 67

   4. Eye and Mouth Cosmetics 69

   5. Skin Cosmetics 69

   6. Aromatic Waters 71

   7. Alcoholates or Tinctures 72

   8. Vinegars 73

   9. Aromatic Baths 74

   10. Emulsions and Milks 75

   11. Pasta and Flour 77

   12. Soaps 78

   13. Oils, Ointments, Glycerols 80

   14. Blushes 81

   15. Powders 83

   16. Sponges 84

   17. Tattoos 85

   18. Perfumes, Medicines and Poisons 87

IV. GENERAL FRAGRANCE APPLICATIONS 92

   1. Perfume Boxes 92

   2. Powders for Sachets 93

   3. Spanish Skin 94

   4. Scented Stationery 94

   5. Scented Bookmarks 95

   6. Scented Stones 95

7. Cassolettes and Spring Dishes 96

8. Scented Shells 96

9. Fragrant Flowers 96

10. Scent Jet Rings 97

11. Scent Jet Pins 98

12. Incense 98

13. The Censer 98

14. Perfume Lamp 100

15. Fumigation Paper 102

16. Vaporizers 103

THE POLITICAL ROLE OF FLOWERS 105

Best sellers Max Milo Editions

Hitler's banker, Jean-François Bouchard

Confessions of a forger, Éric Piedoie Le Tiec

The Koran and the flesh, Ludovic-Mohamed Zahed

Governing by fake news, Jacques Baud

Governing by chaos, Collectif

A political history of food, Paul Ariès

Mad in U.S.A.: The ravages of the "American model",
Michel Desmurget

Mondial soccer club geopolitics, Kévin Veyssière

Putin: Game master?, Jacques Braud

Treatise on the three impostors: Moses, Jesus, Muhammad,
The Spirit of Spinoza

TV Lobotomy, Michel Desmurget